THE BOAT OF FATE

THE BOAT OF FATE

an historical novel

by
Keith Roberts

PRENTICE-HALL, INC., Englewood Cliffs, N.J.

The Boat of Fate by Keith Roberts
First published in the U.S.A. by Prentice-Hall, Inc.,
Englewood Cliffs, N.J., 1974
First published in Great Britain by Hutchinson & Co., Ltd.
© Keith Roberts 1971
All rights reserved. No part of this book may be
reproduced in any form or by any means, except
for the inclusion of brief quotations in a review,
without permission in writing from the publisher.
Printed in the United States of America

10 9 8 7 6 5 4 3 2 1

Library of Congress Cataloging in Publication Data
Roberts, Keith.
   The boat of fate.
    1. Rome—History—Empire, 30 B.C.–476 A.D.—
Fiction. I. Title.
PZ4.R6447B04 [PR6068.015]   823'.9'14   73-20187
ISBN 0-13-077792-7

*For Crearwy
who ate honey*

THE BOAT OF FATE

PART ONE

Chapter one

For many evenings now I have sat patiently, scratching away at a ragged mass of memories. It seems the time has come finally to reduce what I have committed to an order; to give it shape, and a certain logic and progression. Not such a shape, perhaps, as the least of our poets could achieve; I am no author, and make no pretensions along those lines at all. But I have had a full and varied life; I have known despair, terror and joy. Something of that joy I want to set down, to be a spark against the darkness that is surely coming on the world.

Above me the lamp, hanging from its bronze chains, casts a mellow circle of light. Beyond that circle, at its wavering outer edge, hover the ghosts I have raised; beyond them again is the dark, symbol perhaps of that greater Night. Soon, I shall seek my bed; but for a while, who knows? Perhaps the Muse will condescend to rustle her robes a little closer to the Earth.

I am Caius Sergius Paullus. I was born in Hispania in the reign of Valentinian, that Pannonian clown who thought the world was made to benefit barbarians. I was a babe in arms when he died, unlamented, of a stroke; when I was five Theodosius, whom men already call The Great, was raised from his estate in Tarraconensis to be the new saviour of the world.

As one grows older so the easy convictions of childhood become more blurred; and I suppose I of all men should be the last to criticise an Augustus. But I can find little good to say of Valentinian. I've heard he was an efficient administrator; but as far as that goes any Tribune of a Schola, raised suddenly to the Purple, might have done as well. He was a peasant, like his father before him; and nobody is as merciless as a freedman who has happened on authority. He it was who vulgarised the Empire, first decimating the Senate with his purges, then making their numbers up by granting the clarissimate to any of his thick-headed friends who happened to catch his eye. Some of these oafs rose to be Governors of Provinces; and the West laughed, since there was nothing else to do. If the West laughed, the barbarians laughed; and losing their respect for us, also lost their fear. Under Valentinian, the floodgates started to creak; and not even Theodosius could finally stem the tide that burst through.

But it made little difference to us in the far south-west who lorded it in Mediolanum or Rome. In my childhood both Hispania and Gaul were much disturbed, owing forced allegiance to Magnus Maximus, one-time Governor of Britannia and self-declared Augustus. The jockeyings for power among the various factions were complex and endless; we kept our heads down, paid our tax and hoped for better times.

In those days Hispania was still a wealthy Province; indeed Baetica, in which stood my home town of Italica, had been called, like Egypt, the granary of Rome. Our wines were famous, while Hispanian horses fetched good prices anywhere in the Empire, being faster and smoother in action than the celebrated mounts of Persia. It was through the breeding of horses that my mother's family had first risen to eminence, and the status of curiales; so it was no disgrace when my father mingled the blood of Roma with the blood of a Celt.

My mother had a Christian name, a given name, Maria; but it is by her true name, Calgaca, that I choose to remember her. The Bishop of Italica gave it to her, sprinkling her with water and making her swear to follow the true God and his faithful Church. She was very sick at the time, having nearly died in bearing me. My father, like the great Constantine, was content to remain unpurified, having agreed to undergo the ceremony on his death-bed when opportunities for mortal sin were past.

All Hispania was strong for Christ in the days of my youth. Bishops proliferated, so that Baetica alone laid claim to twenty; though as my father once remarked all of them seemed more concerned by heresy within the Church than paganism without. Our hillmen of the north, our Galicians and Cantabrians, still worshipped, as they had always worshipped, what Gods they chose; and some of their methods were most woefully unholy. Meanwhile the Church of Hispania was split by schisms, many of the Bishops rallying to the cause of the zealot Priscillianus; to such good effect that much worthy blood was spilled before that energetic and misguided man was finally executed by the mock Emperor Maximus. Not so much, as my father pointed out, for consorting with lewd women, apparently a human and forgivable aberration, but for presuming to invoke the Almighty without the benefit of his breech-clout; my father, as you will have gathered, was a bitter and sarcastic man.

Like most of the towns of southern Hispania, Italica was surrounded by tolerably high walls. Beyond the gates stretched plantations of olive and orange trees, while the land round about was dotted with the farms and country houses of the wealthy. On market days the little forum would throng with shoppers and mendicants, peasants from the outlying districts with their produce, pedlars of rugs and furnishings, pottery, jewellery, iron and bronzeware, lamps, statuettes. Smiths displayed their wares, waggon- and wheelwrights shouted their skills one against another, perfume-sellers set out their bottles and jars; most things, vendable, wearable or edible, found their way finally to the market-place of Italica.

There were soldiers, too, on leave from duty with our one Hispanian Legion; and not infrequently barbarians in the service of Maximus or Theodosius, proud, stinking men who spoke clacking tongues of their own. They thronged the colonnaded walks surrounding the forum, where the businessmen and officials of the town, my father among them, had their offices; or they would wander across to stare curiously at the long basilica flanking the market-place, the building where once justice had been dispensed and judgements made, long since reconsecrated in the name of the Christian God. Beyond it were more shops, set in narrow side alleys, where drinks, pastry and sticky sweets could be bought or acquired for small services, the running of errands, delivery of goods to customers living outside the walls. Here, too, in this busiest quarter of the town, stood the house in which I was born.

Its entrance was narrow, squeezed and jostled on both sides by the façades of shops. One was a wine-sellers, run by an aged and cranky African who had looked the same as long as I could remember. All day long, and a good part of the night, he would shout his wares in a croaking voice, while his consort, an equally ancient Negress, alternately fussed with the tall jars that held the stock, each with its stopper and leaden vintage seal, or beat a species of doleful accompaniment on the pots and vessels with which one wall of the place was lined. The old fellow loathed children with an abiding hatred; he would drive them from the shop, or the pavement before it, with cuffs and shouts, muttering between times a series of curious oaths taken by most of us to be peculiarly withering spells. My father was his landlord; so with me he adopted a slightly more subtle technique. Were I to pause before his establishment either he or his wife would instantly appear, wielding an enormous broom with which they proceeded to dust the stone flags of the pathway. I would be shooed aside, rather like a fly brushed from a joint of meat, and, unless I skipped pretty smartly, with a well-barked shin or two for my pains. In time I learned to avoid both the virago and her master, contenting myself with taunts muttered at what I deemed a safe and suitable distance.

On the other side of the porch was a bakery run by a Syrian, a fat, wheedling, olive-skinned man with a high-pitched voice and a truly remarkable squint. The legitimate sale of bread and pastry formed only a small part of his income, if the gossip of the town was to be believed; he was a notable procurer, both of young women and boys, and despite the undoubted value of his services he was heartily disliked in Italica. Some of his team of young acrobats worked in the shop, sweating over the ovens or grinding away at the big cornmills that stood against the rear wall; their rumbling penetrated the house, reaching to the atrium and beyond. However, Septimius, as he styled himself, usually seemed anxious to keep on good terms with me; most times, if he saw me passing, he would waddle from the shop, press on me a bag containing pastries, piping hot and fresh, twirled sometimes into shapes like curious little mannikins. He was frequently to be seen about the town and market-place, stertorously taking the air or unobtrusively touting for trade.

My father's house, though small, was well proportioned and excellently maintained. Like all our Roman houses, except the homes of the very poor, it was externally featureless. The doorway was narrow, as I have said, and closed by two heavy bronze-studded valves that creaked abominably on their pivots as they turned. Beyond, the place was of classical design, its rooms and offices grouped round the open-roofed areas of atrium and peristyle. The atrium was curious in that there was no central catchment tank for rainwater; instead the roof-eaves sloped outwards, the water draining through hidden conduits to an underground storage system. Though undeniably less elegant, my father considered the arrangement more hygienic; he set great store by such things, and was always concerned to see the housepeople were kept up to his own exacting standards.

Sleeping cubicles, each paved traditionally with white mosaic, opened from this forecourt of the house, while to one side stood my father's massive iron-bound safe. It contained little of real value; just some family papers, and a couple of dusty and much battered masks of ancestors. I remember as a child how terrified I was of them. On feast days, according to ancient custom, they were taken out, furbished and displayed; I would skirt the alcoves in which they were placed, if possible with eyes averted. Looking back there seems little to have inspired such fear; one represented a stern-faced, beak-nosed man with a squint not unlike that of old Septimius, the other was of a woman with an odd piled-up hairstyle and a fatuous, rather wooden grin. In fact one of the housepeople—probably Marcus—muttered once that in his opinion they had nothing to do with the history of the family but had been picked up as a job lot at some auction of the trappings of a temple. I am inclined to believe they were genuine, whoever they represented, for my father was moved to quite extraordinary rage when the tale was carried to him. Soon after that the custom of displaying them was discontinued anyway, and I never saw them again; for the good Bishop of Italica at last condemned the practice as idolatrous, and my father, as a man very much in the public eye, was forced, grumbling bitterly, to defer to the strictures of the Church.

Beyond the atrium an opening, wide and high and generally closed by richly decorated drapes, gave on to the garden of the peristilium. Here roses and lilies stood in carefully ordered rows; and here too were many fountains, filling the place with their lilt and splashing chuckle. Water supply was my father's profession, but it was also his hobby. If complimented on his garden he would usually remark that though money might have vanished from the scheme of things, fountains at least were free.

A colonnade ran round the peristyle, giving shelter on all sides from rain and the direct heat of the sun. More rooms, my father's study and the family sleeping cubicles, opened from it; beyond it the triclinium, a long narrow chamber flanked by its attendant kitchens, terminated the house.

Before the screen wall of the triclinium, placed on a pedestal where it could watch back through the house, stood a most curious bust. My mother commissioned it, from a wandering Ceard; a slim, corn-brown man who laughed and chattered as he worked in the singing, lisping tongue she would never use. Though as he spoke she smiled and nodded, and once the tears came suddenly to her eyes and she brushed them away and smiled again like sun after rain. I was very small at the time; I crouched at her knee as the stranger tapped and chipped, darting forward now and again to catch the fragments of stone as they fell or were brushed from his work. I was ranging them on the flags of the courtyard, pretending they were armies of Romans and Persians drawn up to do bloody battle, for I was already imbued with a sense of the history and destiny of my father's people.

The statue was intended as a likeness of that Virgin who the Bishop assured us had been the Mother of God; but the strange eyes, mere holes above the gash of a mouth and the half-existent nose, the full face and profiles separate yet curiously joined, were like nothing I or the household had ever seen before. My father detested the carving, though he suffered it to remain. The housepeople were afraid of it. Even Marcus shook his head over it and muttered that he had seen its like only once before and that on the Wall, the great Wall the Emperor Hadrian built to mark where the last Standards were to rest, on the grim edge of the world. Which was curious, for I learned afterwards the craftsman had indeed hailed from Britannia.

My earliest tutor was a Greek, from Alexandria. He remembered, as I suppose a Greek well might, the man most Romans had chosen to forget. His bust (and a statue of the Divine Antinoüs, who Heraclites told me was still worshipped in Egypt, where he met his end) adorned the little room, half study, half living-quarters, where I took my first lessons. Publius Aelius Hadrianus, builder of walls, who secured for the Empire a century of peace; the Antonine succession, blighted finally by the Divine Commodus. Hadrian himself, of course, was an Hispanian; more than that, he had been born in my own town, in the great white villa on the outskirts of Italica still owned, if their tale was to be believed, by his descendants. He had trodden the streets I walked, perhaps even thought the thoughts that at times thronged my brain. I would stare at his statue by the hour, in between wrestling with the outmoded Greek Heraclites would insist I learn; sometimes, left on my own, I would draw a footstool close, climb up to gaze into the Emperor's dusty blind painted eyes. At such times it seemed the great soldier spoke to me; though only once he visited me in a dream.

My mother set great store by dreams; for once, when herself was a child, she had seen a famous King with his winding sheet and coffin beside him, and what she prophesied had come to pass. So when I saw the night-shape I was more than usually ready to be impressed. In the dream I found my voice, asked quaveringly then more firmly, 'Who are you, sir?' And the Shape—it was little more—laughed musically, though its voice when it answered was gentle and sad. 'I am he whom you seek,' it said. Then, cunningly, 'The man from the Adriatic. . . .'

At that a great joy seized me. I held my arms out, tears starting from my eyes; and the vision was gone, I lay in a room bright with moonlight and rich with the scents of flowers, hearing a night-bird cry and the echoes of my own wail. Till Calgaca came, and soothed; for she would suffer no serving maid to attend a child in nightmare. I remember the rustle of her robe, the strength and firm coolness of her arms. She brought me water, wiped my face and forehead, rearranged the covers of the bed; she sat with me till the lamp burned dim, its flame browning and flickering, and I fell into a dreamless sleep.

Such night alarms were not infrequent, for I was a nervous and excitable child. I think this was largely due to the remorseless faith of the housepeople in all things supernatural. The Christians, of course, had damned the whole pantheon of Roman Gods, translating their virtues and follies alike into devilish characteristics; and incidentally exhibiting more faith in the traditional deities of the Empire than the average enlightened Roman had shown for generations. In addition the rocks and soil of Hispania had produced their own rich crop of horrors; and nobody was better versed in such folklore than my old nurse Ursula. She was an extraordinary woman, tall and gaunt to the point of emaciation, with frizzy hair, an ugly, gentle face and the longest, flattest and boniest feet I think I have ever seen. Ursula was three-quarters Celt, and I'm inclined to think more than three-quarters mad. It was she who regaled me, usually at my own prompting, with tales of lemures and werewolves, vampires eating the noses of dead men; and there were old ladies who changed with the dusk into taloned birds, and a great Man who haunted the northern seas, reaching with his scaly arms to overturn the stoutest ship. So it came about, not unnaturally, that nursery bogies plagued me; Morphio with her loping donkey's legs or Lamia, her belly swollen with a screaming, burning child. At such times Calgaca's eyes would flash, the colour deepen in her cheeks; she would hold me against her breast and talk of strange people and plants and animals on the rim of the world till my eyelids drooped and the curtains stopped their stirring and whispering and the house was quiet. Sometimes she would send for Ursula, vowing to dismiss her from her service for filling my head with such unpleasant rubbish. Then the strange creature would snuffle and weep, clinging to my mother's robe and swearing she would die before she injured me; and Calgaca would relent and dismiss her to the kitchen, where it took several draughts of unmixed local wine to restore her nerves. By which time, of course, she was usually in a condition to start again. I think, simple soul that she was, the old woman truly loved me; and those half-delicious terrors did no lasting harm.

My father had been born in Rome, to a middle-class equestrian family, though he emigrated to Hispania as a young man and remained there the rest of his life. Why, I was never too sure; except that maybe he preferred to be headman of a village rather than second-in-command of an Empire. He held an important position in Baetica, as a Curator in charge of the public water supply. His duties frequently took him away from the town, so that during my early years I saw very little of him; if he was at home he would almost invariably be shut away in his study at the far end of the house. When he was working there at his ledgers and files, or reading in his extensive library, it was more than either my mother or the servants cared to do to interrupt him. He was a stocky, quietly spoken man, balding and with steady, piercing eyes; and that truly was almost all I knew of him. About the house he was always formally polite, although he was prone to bouts of bad temper, when he would curse with a violence that was truly terrifying. He was, before anything else, a figure of authority, and I was never able to feel really close to him. In any case, I was very little interested in his work, though I learned afterwards to value his skills more highly.

At rare intervals, when the mood took him, he would call me aside from my studies. Then as like as not he would endeavour to drive into my head some notion of the complexities behind this simple matter of piped water. His staff was divided into three sections, each of which was kept to strength, in theory at least, by levies supplied by the State; though with the constant demands of the Army taking priority manpower was usually woefully short. Also, of course, the edicts of successive Emperors binding humiliores ever more closely to their family trades had rendered free recruitment almost impossible, though there were men enough in Italica alone who would have been glad of the chance to draw the rations levied against the countryfolk in return for Government employment. But such requests might have to pass to the office of the Praefect of the Gauls himself, where after years in limbo they were likely enough to be turned down; and my father was reduced to bending the law, with the active assistance of the local authorities, none of whom fancied existence without those public services to which a lifetime as citizens of the Empire had accustomed them. Thus miners and peasants, as experts in handling soil, might be set to digging out blocked or damaged mains, stonemasons worked on clearing birds' nests from the air vents of aqueducts; the labour shortage was largely overcome by such evasions, though the increased paperwork and the endless need for the invention of fresh euphemisms added lines to my father's forehead, and a cutting edge to his naturally acrid tongue.

One of the motley labour forces thus formed travelled continuously repairing the channels of aqueducts, renewing the lining slabs where the endless rushing of water had worn them thin; another section maintained the fabric of the great arches that were built wherever the channels were forced, in their carefully-devised routes, to cross a valley; while a third gang was often to be seen in the streets of Italica itself, attending to the mains that ran beneath the pavements.

Where they entered towns the channels flowed into massive

towers within which systems of overflowing tanks distributed water for domestic and public supply. In times of drought these regulating devices ensured that the street fountains were the first to cease to work; after that the public baths ran dry and finally it was the turn of private householders, most of whom were hard put to it to get by without their constant supply of piped water. In the south of the Province severe droughts were infrequent, though once the town baths of Italica were closed for several weeks. Everybody was surly and bad-tempered till the supply was restored; my father was surlier than the rest, having been called before the duovirs and formally admonished for his part in the general inconvenience. At such times he was apt to remember his breeding, muttering that it was no part of the function of a Roman to answer to a pack of indigenous tribesmen. There was in any case always a certain amount of bad blood between Government officials, who were exempt from curial duties, and the unfortunate town senates, who since Valentinian's confiscation of the city taxes had been forced to dig deeper and deeper into their own pockets to meet their responsibilities.

Somewhat similar factors tended to sour my father's relations with the Church. Members of the clergy were likewise free of the burden of the Curia; hard-pressed gentlemen in danger of civic responsibility still managed to acquire Holy Orders without divesting themselves of their estates, despite the array of legislation aimed at curbing the abuse. I remember the delight with which my father greeted Theodosius' fulmination against such ordained curiales. He had been very much on his dignity with the local Bishop, who he suspected of having acquired office for reasons far removed from altruism; for weeks after the injunction was published he would stop him in the street to enquire gravely after his health, and urge him to pray for the soul of an Emperor so patently in need of spiritual enlightenment.

The row in the Curia acted indirectly in my favour; for when the water-level was finally restored, my father, to ease his smarting dignity, set out on a tour of the district under his control, which he afterwards extended to take in the domains of his associates, some of whom he had not seen for years. For once I was allowed to go with him, and saw a great deal of the Province for the first time. We travelled finally to Segovia to view the great aqueduct there, which has the finest and longest span of arches in Hispania. I still remember my first glimpse of them, towers of stone blazing white under the brilliant sun, striding like the legs of giants into farthest distances.

Such occasional outings were the highlights of my life. Sometimes I travelled with my mother; I remember particularly one visit we made to a race meeting in which my uncle was running several teams. I came home with a burning resolve, mercifully short-lived, to be a charioteer myself. Even today I can re-create that journey in my mind; the lurching of the carriage as its wheels crashed into pot-holes and bumped over ridges, the smell of horses and leather, the swirl and white billowing of the all-pervasive dust. And clearest of all my mother's eyes, dark blue and lovely in the bright sunlight, as she listened gravely, despite the amusement she must have felt, to my grandiose plans.

If my mother was one of the great influences of my life, Marcus was another. Old Marcus—I always thought of him as immensely ancient, though of course he was not—was a time-expired Legionary, a man who had seen service on half the borders of the Empire. His family hailed from the distant Province of Noricum, north of the Adriatic Sea. His father had been a schoolmaster and teacher of rhetoric, a poor man who to support his large family had been forced to apply for grants from the State. Aid had been given, but at a price. Marcus' ambition had been to become an advocate; he had studied hard, and still retained a smattering of Greek. But at eighteen the Army claimed him; he was posted to Divitia, to the headquarters of the Second Italicans, to learn to be a soldier. His intelligence earned him rapid promotion; within ten years he had attained the rank of Tribune and was serving in Gaul, with the comitatus of the Caesar Julian.

Marcus had never spent his Gallic donative. He showed me the coins one day, gleaming solidi of the Emperor Constantine; he kept them in a strong-box tethered to the wall of his room. He served my father as doorkeeper and general handyman. He had a little cubicle set next to the atrium, equipped in part as a workshop; you could rely on finding him there most times of the day, busy with the repair of domestic implements, harness, shoes. He was a man of most varied abilities, for he was also principal gardener to the household, and what he planted invariably seemed to thrive.

In addition, he kept ferrets. He bought them from a Bithynian dealer who occasionally passed through the district; they were imported from Libya in considerable quantities to keep in check the little burrowing hares the countryfolk called peelers. At one time the bloodthirsty creatures lived in a range of hutches in the peristyle; but their stink eventually came to pervade the whole house and my father banished them to the stables a few yards along the street where he kept his horses. I learned early on to handle them, and was seldom bitten; one of my chief delights was to be allowed to ride out with Marcus, usually in the early morning, on his hunting expeditions. On a good day we would ride back to breakfast with our saddles hung with bulging sacks of game. The meat was a welcome addition to our household supplies, for it was often difficult for my father to collect the ration allowances due to him, while money was scarce and rapidly becoming scarcer. But what else could you expect, he would mutter, when the taxes levied from the country went it seemed to fight half the barbarians in the world?

Usually when my lessons were over for the day I would slip through the atrium to Marcus' workshop, sit and swing my legs on the bench and plague him for tales of Goths and Vandals and Alamanni, Persian Kings with silk and jewels and retinues of bare black slaves. Some of his best tales were of the vast, empty lands to the East; for he had followed Julian on his ill-fated expedition into Persia, fought in the great battle in front of Ctesiphon. The arms of the Empire were victorious; but no Roman ever entered that huge town. Instead the army swung unwillingly north, led by its pale-faced Emperor; for Julian the pagan had taken auguries, seen Death in the scattered guts and bones. 'His mind wasn't on it,' Marcus would say. 'You could tell . . .' And his hands would still for a moment, his eyes narrow as he stared beyond the confines of the little room. Seeing that vast army crawl day after day across the barren land; hearing the rumble of waggon wheels, the scuff and weary tramp of feet; and over and again the drum-roll of hooves, the cries of fear and rage and clatter of arms as the Persian cavalry swooped on flanks and rear. Man after man, Roman and barbarian alike, thudded to the ground, gasped his life out on the harsh earth; cairn after cairn was built, left for the jackals and buzzards and the endless droning wind. Till it seemed, to weary soldiers and a wearier King, that they were all cursed, that they would wander in a trackless waste for the rest of time.

For one of them such prophecy came true. Marcus described it well; the bleak rumour running through the camp, from cohort to cohort, century to century. Julianus Augustus, ruler of the world, was no more. Once again the Empire was leaderless.

'That was a time, my lad,' Marcus would say. 'That was a time, indeed . . .'

They picked another leader for themselves, out in the wastelands. 'And why they chose him,' said Marcus, 'only the Gods can say. I shall never know.' Jovianus, a well-meaning ass from the Corps of Protectores, signed away at a stroke the satrapies across the Tigris owned by Rome since the days of Diocletian; eight months later Jovianus too was gone. 'They say it was natural causes,' said Marcus grimly. 'But I wouldn't bank on that. There were enough at staff headquarters ready to spice his soup . . .' The army, dispirited, made for home; and Valentinian, of whom I've already said more than enough, was finally elected with his brother at Nicaea. The pair proved little better; and had it not been for Theodosius, a modest landowner of Hispania, the Empire would have been laid in ruins there and then.

Marcus marched with his father, the Magister Militum who restored order to the restless Province of Britannia; he ended his service there, and married a Britannic wife. He was offered a sinecure, as a camp commander in charge of training limitanei, but he declined: he'd seen enough of soldiering and war. As gratuity he was given a house and allotment in the garrison town of Eburacum; five years later his wife was dead, and Marcus traded his land for mules and a waggon and headed south, crossing Gaul by slow stages to Hispania where he took service with my father. Behind him, Gaul and Britannia burst into rebellion again under Magnus Maximus; he shrugged it off, sitting in my father's hall, stitching and hammering for his keep. 'All I got from Rome,' he would say, pointing, 'is in that bloody box. That and a patch of cabbages, that weren't worth the sweat I dropped on them. . . .'

But he was seldom bitter, at least with me; he taught me to throw the dice, the ivory cube and oblong every soldier carries, swig the sour wine he still insisted on drinking. My mother only half approved of his hold over me. Certainly it was from him I first caught my military enthusiasm, and formed the patently absurd resolve to be a soldier. Mastery of the Army, as Marcus himself pointed out, had long since been given over to barbarians. Duces, Magistri, Comites, all came now of Frankish or Scythian stock; and such men invariably promoted their relatives and tribespeople, over the heads of Romans bred and born. In any case, as the son of a Government official the armed forces were denied me. My father had already determined I should study law, from which beginning I might rise to be a civil administrator, a Praeses, Praefectus or even a Vicus in charge of a vast Diocese, for such posts were frequently given to candidates of equestrian rank.

All this meant nothing to me. Through my dreams ran a remorseless, endless thunder; the marching tramp of Legions. They poured past in majestic array, sunlight glinting from harness and shield and helmet, sword-hilt and javelin-point. Sometimes I myself was at their head; I rode a gilded chariot through the streets of Rome, a triumphator come to claim the honours that were his due. What Hadrian had done, I had done as well; secured the Empire, restored its ancient glory. Before me strode the Senate, resplendent in their official robes, the spoils and crowns of victory borne high in their train. Trumpeters cleared the way through the roaring mob; there were white bulls for sacrifice, and chanting priests to guard them; lictors with their wreathed bundles of rods, prisoners trudging bowed down in their chains. I stood stiffly in the chariot, bracing myself as its wheels crashed over the rutted flags, eyes staring straight ahead, face glowing behind its mask of scarlet paint. Folk I knew, the housepeople, Marcus, my own father and mother, cheered me from the streets. And the cocks would crow, across the roof-tops of Italica; I would sit up panting, seeing my own room in the first cool light of dawn.

I won no glories, at the age of nine. Instead, I went to school.

Chapter two

The change in my fortunes was abrupt; too abrupt by far for my taste. My father announced, one night at dinner, that the household could no longer afford the cost of a private tutor, and that in any case it would do me good to get out and about and mix more with my fellows. I was sent for the next morning and informed brusquely of my fate. My protests fell on deaf ears; I was dismissed with equal sharpness, and scurried to find Marcus. To my disgust, he was all in favour of the notion. I did extract a promise from him that if I made good progress with my books he would, with my father's permission, take me hunting deer; and with that I had to be content. I even applied to Heraclites for sympathy; for though I had cursed him often enough, when he kept me at my books in spite of the sunlight pouring into the house, I had taken my first lessons from him and the Devil one knows is always preferable to Devils of the imagination. I, who had already in my mind conquered the world, felt a sharp reluctance to leave the four walls of my home.

The secondary school in Italica was run by one Gellius, a tall, lean and embittered Grammaticus whom I had seen often enough striding through the streets on his way to or from his place of torture. He was a bad man to cross, by all reports; free with his tongue, freer with the strap. However, my fate was obviously sealed; I set about packing my writing materials, pens and cases, paper, wax tablets and the rest, and started out on my first morning prepared to meet my doom, if need be, with all the stoic determination of my forefathers.

I think freedom is a condition only truly appreciated after it has once been lost. Certainly I had never realised until I was packed off to school what a comparatively untrammelled existence I had led. Gellius held court in rented premises above a laundry a few streets from my home. The room in which we worked was low-ceilinged and gloomy; fusty in winter, in summer unbearably hot. Diminutive windows, set close under the eaves, admitted little more than stray breaths of air, while to mitigate the sounds of commerce from below the door was always kept firmly closed. Gellius sat at one end of the chamber on a heavy, straight-backed chair; the rest of us scratched away at a series of low benches, their surfaces deeply carved with initials and monograms. I generally tried to secure a place near the back of the class; for I was an inveterate dreamer, able to lose myself by the hour in watching the slow swirl of dust motes in the stray beams of sunlight that penetrated the place. That same sunlight lay across Italica; I would see, in my mind's eye, the lush water-meadows beyond the town through which the Baetica flowed on its way to the sea, the lagoons and reedy marshes where the beavers made their homes. Till now, watching them had been one of my constant pleasures; if you sat still long enough, hidden among the tall, mysteriously rustling grasses, the little creatures would learn to ignore you, busying themselves with their complex works of engineering as though nobody was near. Now that pastime, along with many others, was denied me; instead I had to listen to Gellius' rasping voice, puzzle my head over his endless, stupid exercises. The contrast was nearly too much to bear.

I suppose it was natural that I should soon fall foul of my new instructor. Gellius began to single me out, hurling awkward questions at me when he knew my attention had wandered, taking sardonic pleasure in picking what answers I managed to make to pieces. Nothing I had learned from Heraclites seemed to be correct; my mathematics, spelling and shorthand were appalling, while my classical knowledge was dubious in the extreme. I was detained night after night, to the unconcealed amusement of the rest of the class, while Gellius tried to din into me what he considered the elements of a sound education. Heraclites looked in from time to time to check on my progress; he did some work with Gellius, mainly in the higher sphere of rhetoric, in addition to his round of private tutorials. At first I was glad to see him, but I soon learned to dread his visits. The two savants were bound to disagree, usually violently and invariably on some minor point of principle; I suffered regardless of the outcome of these battles, as I remained a captive spectator throughout.

Undoubtedly, though, the worst aspect of the school was the stench from downstairs; for as well as cleaning clothes the shop undertook dyeing by the Tyrian method, in which large quantities of urine are employed to fix the colour in the cloth. To this end stone tanks were placed outside in the street, passers-by being invited to contribute to stocks; in hot weather the vats stank abominably, the reek permeating the whole building. In addition a steady din rose from the place, the thudding and banging of clothes pounded in stone troughs, never-ending screeching of washerwomen, the clatter and grumble of a variety of machines; it all formed a weird accompaniment to the golden verses of Virgil, the meticulous phrasing of Quintilian. Under the circumstances I find it remarkable that I retained any feeling at all for our national heritage; for to this day, when confronted by one of Gellius' favourite passages of Horace, there seems to rise from the pages not the glorious stream of images the poet would no doubt have me see, but that awesome reek of piss.

As I came to know my subjects better, and more importantly to understand the whims of my new instructor, I found life easier and began in fact to discover compensations. Gellius was of a strictly classical turn of mind; he only tolerated Quintilian as a fellow Ciceronian and would have nothing to do with the work of the moderns, those Christian poets who had sprung up, as he put it, like pale weeds among fields of Roman wheat. In this at least I could follow him, Heraclites having been of very similar temper. I remember once the local Bishop, no doubt to improve our collective minds, loaned us a tract by Prudentius, a Tarraconensian much in vogue at the time. Gellius, in a mood of bitter humour, set us to extol the poet's virtues as an exercise in rhetoric; I damned him instead, to my tutor's ill-concealed delight, ending my address with the pious hope that he might one day be compelled to lick clean all the parchment he had spoiled. My speech, delivered stumblingly and in trepidation, proved the turning point of our relationship. Praise from Gellius was praise indeed, mainly because of its scarcity; I applied myself more enthusiastically to my studies, coming eventually to be considered something of a teacher's pet. The envy of my schoolfellows meant very little to me, for despite my father's hopes I had made no close friends of my own age. There seemed to be a queer, lonely streak in me; as an only child I had, of course, been forced back to a large extent on my own company, and I looked for nothing better now.

In the main, though, I got on well enough with the rest of my class, with one exception: Publius Aelius, as he styled himself. His parents owned the big Villa of Hadrian just outside the town, a fact he never allowed to be forgotten, though despite his airs he attended the same school as the rest of us. He was my elder, I suppose, by a couple of years, a tall, olive-skinned boy with black curly hair and smallish, restless dark eyes. From the first his attitude towards me had been condescending in the extreme; 'Poor Sergius' and 'Young Sergius' were phrases never far from his lips. I bore his ways without too much difficulty, as I was frankly unable to take his bombast seriously. As I rose in my tutor's esteem, however, his attacks became more overt; till one day I realised, belatedly, that a dangerous situation had developed.

Gellius' school breaking up for the day always produced an effect like the collapse of a small dam. Once released from our place of torture, the whole score of us would pour down the narrow, smelly stairs and race in a mob down the alley that ran alongside the building, each determined for some obscure reason to be first into the street. On this occasion somebody elbowed me heavily in the rush, knocking me off balance and sending me crashing into Publius. The books he was carrying flew in several directions, one ending up with a splash in a muddy puddle; next instant I had received a box on the ears that made my head ring. I straightened, eyes stinging and fists clenched. Publius had flushed a dark, brownish red. For a moment I thought he would fling himself at me; then he relaxed, turning away with a sneer. 'Save your effort, young Sergius,' he said. 'When I deal with you, it won't be with my hands.' He stooped sullenly, began gathering up his scattered belongings.

I said angrily, 'What do you mean? I'm sorry about the books. It was an accident.'

He looked up, his eyes malevolent. Between them projected what would in future years be a haughty beak of a nose. 'You should be more careful,' he said. 'My father is already teaching me the use of the sword.' He wiped a spoiled book on his tunic. 'We come from a long line of soldiers,' he said. 'You are the son of a freedman and a runaway Celtic slave. I am descended from Hadrian.'

Despite my annoyance my brain worked quickly. Some of my father's bitterest gibes had been directed at Publius' family; he told them to Heraclites, who lost no time passing them on to me. 'That would be difficult,' I said, 'seeing the Empress Sabrina habitually boasted of her infertility. Perhaps that relative of hers had something to do with you, though, between writing those lousy verses on the Sphinx.'

The sally, though geographically inaccurate, brought him to his feet. I thought for an instant he was going to attack me again, and stiffened to meet the onslaught; then he laughed, a little too loudly to be convincing. 'Poor Sergius,' he said. 'I keep forgetting. You are still only a child, who cannot yet handle a sword without pricking himself. I don't fight children.' He turned on his heel, walked away with some dignity and was gone.

I made my way home in a somewhat chastened mood. I was still unsure how the situation had come about, but it seemed I had made a thoroughly unpleasant enemy, and some time in the future would have to play my part or bear the consequences of cowardice. Also my ear and cheek still burned a little; and when I thought back to Publius' insults I felt my temper rise again. That night I sought Marcus in his room. I mooned about for a while, fiddling restlessly with this and that, till finally he told me curtly to find something useful to do or clear off and leave him in peace. I swallowed at that and said suddenly, 'Marcus, will you make me a sword?'

He put his work down, stared at me gravely in the lamplight. He said, 'And why would you want a sword?'

I said primly, 'To defend my honour.'

Marcus smiled at that, slowly. He said finally, 'What's wrong with your fists?'

I said, 'I think it's gone rather too far for that.'

He pursed his lips, frowned, whistled between his teeth. Then he rose and walked towards me, stood arms folded looking down. He said slowly, 'Caius, a sword is not a toy. Nor is it just a weapon. It's like a badge, or the Standard a Legion carries in front of it. It tells other people something about you. It tells them you are no longer a boy, that you're no longer looking for concessions. It tells them you think of yourself as a man, that you're prepared to live in their world and bide by its rules. Once you take it in your hand, there's no more backing down. You have to finish what you have begun.'

I felt rather annoyed at him. He seemed to be criticising me, though I wasn't sure how or why. I said, 'I have thought about all this, Marcus, and my mind is made up. I want a sword.'

He still waited, frowning, rubbing his lower lip with his finger. He said, 'And would your mother approve of what you want to do?'

I said quickly, 'She doesn't know. You mustn't tell her. Promise, Marcus.' I hesitated. I said, 'It's partly for her sake anyway.'

He shook his head at that. He said, 'Caius, Caius . . .' He turned away, abruptly; then he swung back. His face looked queer; angry, and dark. He said, 'Very well. A sword you want, a sword you shall have. Come to me tomorrow evening. Wear an old tunic; I shall want you to work the bellows.'

He was as good as his word. I lit the forge and pumped while he banged and hammered, striking showers of sparks from a strip of glowing iron. Later, working under his direction with needle and waxed thread, I fashioned a scabbard for myself and a belt from which to hang it. Marcus made a hilt and cross-piece, tempered the blade and scoured it with sand; and finally, a few nights later, the work was done. I took the thing in my hand; a little bright heavy blade, tailored to my grip. I could scarcely believe it was mine.

'Take it away,' said Marcus gruffly. 'Don't wave it about in here. And mind you learn how to use it.'

At school things were getting better and better. Gellius, having detected at least the stirrings of a fellow spirit, thawed considerably, devoting more time to me than anybody else. The evening sessions still went on; but now they became a pleasure. The concentrated training in rhetoric improved both my diction and my memory; I bored and baffled Marcus by turns with classical dissertations till eventually the long-promised deerhunt took place.

It was an exciting affair. For weeks beforehand Marcus had me collect large feathers, as many as I could find. What use he would put them to he refused to say. We camped away from home for several nights, in a belt of woodland half a day's ride north of Italica, where Marcus spent some time preparing a complicated and ingenious trap. The feathers, many of them dipped in bright dyes, were plaited into light ropes that he strung from a series of posts set between the trees. A tweak at one of them would set the feathers twirling and spinning for yards; deer, he assured me, would not cross the barriers, though they presented no real obstacle. The trap was funnel-shaped when he had finished it, a hundred yards or so across the mouth and tapering to a narrow corridor, in which he dug several pits. These he camouflaged with branches and dried grass; then came the ticklish business of the drive. He rounded up a handful of local peasantry, all of whom were more than ready to co-operate on the promise of fresh venison. The deer, pretty, dappled creatures with dark, sad eyes, were herded finally into the trap, where they were finished off brutally enough with clubs and stones. The last part of the process I found I couldn't watch; I still had a lot to learn before I could become a soldier.

A few weeks afterwards my growing scholastic prowess came to my father's ears. Gellius, like most teachers of rhetoric, was in the habit of holding public demonstrations, at many of which I had become his star attraction. On this particular occasion my exercise was to argue on Hadrian's side in justification of his war against the Jews. It was a subject close to my heart; I had prepared my material well, and my impassioned pleading brought the house down. A couple of evenings later my father sent for me. He was sitting writing when I entered, and for some minutes continued to scratch away without acknowledging my presence. When he finally spoke I had a great surprise. He gave me, for my own, part of the *De Re Rustica* of Columella.

The gift was totally unexpected. I flushed with pleasure, and began to stammer out my thanks, but Father cut me short. It was then I first really sensed the gulf that existed between us. 'Don't misunderstand me,' he said in his quiet, cold voice. 'I'm rewarding you for your accomplishment as an orator, not your opinions as an historian. You speak very well; but for the rest, you deserve a thorough whipping.'

I flushed again, this time with annoyance, and would no doubt have embarked on a justification of my ideas if he hadn't stopped me with a peremptory wave of his hand. 'That's enough,' he said. 'Go back to your room, Sergius. Don't put your orator's scarf on in here.'

The incident had one long-term result. From then on I was allowed the use of my father's library on rare occasions, usually when he was away from home. His collection was extensive and he was himself the author of several treatises on public engineering as well as a history of Baetica that had enjoyed considerable popularity some years before. One wall of his study was lined with the shelves on which the books were kept, each in its wrapper of dyed parchment; to one side was his writing table, illuminated at night by a pendant many-spouted lamp. The air of the room was heavy with a combination of scents: gum, resin, the soot and wine dregs from which he made his ink, the cedar oil used to protect papyri from the ravages of insects. It was a pleasure to me, on those odd visits, just to take the books down and handle them, rolling and unrolling them on their rods of wood and bone, admiring the neatness of the columns of script, the edges of the leaves smoothed with pumice and stained with bright dyes. I glutted myself on the history of the Province; military history for the most part, the campaigns of Gnaeus Pompeius, Sulla and Aemilius Paullus, who I privately believed to be my ancestor. I read his great speech from the Roman Rostra, and Hadrian's exhortation at Tarraco. One day I would visit these places, walk in the footsteps of these famous men; even perhaps—for my ambition had revived—surpass them.

It was about this time that an incident took place that, looking back, I can see was to shape my entire life.

It happened on a bright day in early summer. I was on holiday, and had risen early hoping to persuade Marcus out on a ferreting expedition. But for once he was unco-operative; a great deal needed doing about the house and in any case, he said, an old wound was troubling him, a spear-thrust in the leg he'd received in the Persian war. He shooed me determinedly from his room, and I was left to my own resources.

The day was too fine to waste in reading or study. I considered going round to the stables to try to persuade my father's groom to let me have one of the ponies, but it would have been a waste of time; I was still not considered old enough to stray too far on my own and Victor was under strict orders to that effect. My mother was in the kitchen supervising the preparation of a meal; the rich scent of boiling grape-juice mixed with the sharper tang of liquamen reducing in its earthen pots told me there would be visitors that night. I would not be missed for an hour or so; I slipped out by the servants' entrance, gaining the alley that ran alongside the house. I paused long enough to bring old Zenobia from her den, clutching her enormous broom; my father had christened her that, after the fiery Queen of Palmyra who once caused a slight embarrassment to Roman arms. Then I set off jauntily for the market. At my hip, bumping my thigh as I walked, swung the little sword. I jogtrotted importantly, feeling the drag and bounce of the little weapon, very much a man.

As usual there was a great deal to see. Most of the shopkeepers were busy already, setting out their wares in stacks and piles; in the Street of the Wine-sellers a few early-morning drinkers lounged, cups in their hands, watching me incuriously as I passed. In the market I dodged round one of our housepeople, bargaining for a barrel of oysters. I ran on down the narrow street beyond, where the poorer folk and artisans lived, to the town wall and the west gate. It was hardly ever guarded; I passed through, dropping to a walk, followed the white paved road to where the first of the olive plantations began. Beyond, a mile or so from Italica, was a patch of uncultivated land, thick with saplings and tall weeds. I had played there often enough before; I could normally rely on being undisturbed.

None the less, the wood was full of enemies. The saplings were Goths and Alans, the weeds lurking Persians. I fell on them all, with wild war-whoops; not all of them in my native tongue, for years back Marcus had taught me the battle-cries of the German and British auxiliaries. The Persians fell most satisfyingly; the Goths were more resilient. In fact one of them resisted the sword so successfully as to cause a major burr in its edge. I was disappointed; I would have to wait now to catch Marcus in a good mood, and persuade him to beat it out for me.

I was tired by the time I reached the far end of the copse, and streaked with sweat. I rolled in the grass, lay face down, feeling the blood pound in my temples. I wondered whether to go back into town to the baths, and dismissed the idea. I would almost certainly be caught and scolded by one of the servants, and my day out spoiled. I got up and walked on again, to where the white walls of the Villa of Hadrian cast a pleasant shade.

It stood a little way back from the main road that ran west to Lusitania. It was an unusually fine house, built round three sides of a square; the fourth side, facing the road, consisted mainly of stables and buildings that housed farm implements and waggons. I hung about round the gateway, peering through at the neatly tended courtyard, till a surly slave shooed me off. I wandered on again, round the windowless walls of the place; beneath them, I was safe from observation.

Behind the house, some yards from its rear wall, stood a tall clump of trees. I lay down at the foot of one of them, on my back in the grass, holding the sword above me. I juggled with it, feeling how the weight balanced in its pointed tip bent my wrist forward and back. After a while I half closed my eyes, lay watching the odd clouds that crossed the deep blue sky. If I concentrated I could make it seem as if the clouds stood still and it was I, and the solid earth beneath me, that moved, bowling along majestically to an unknown destination. In time the insect-hum round me faded, and I began to doze.

I was roused by a hard kick on the shin. I sat up resentfully, still half asleep, shielding my eyes with my hand. Publius was standing over me. He was wearing a tunic of fine yellow linen; and his pointed-chinned face was alight with suspicion and dislike. 'What are you doing here?' he asked arrogantly. 'This is my father's house, and this is private land. You have no business on it.'

I was annoyed by his attitude, and more by the kick on the shin; but I had no real wish to give offence. 'I was playing,' I said in a conciliatory tone. 'I came through the wood.'

He ignored me. He had seen the sword that lay at my side; he smirked, and shoved it contemptuously with his foot.

'What is this?'

'You can see very well what it is,' I said, getting angrier. 'A sword.'

'I see a toy,' he said. He stooped quickly and snatched up the sword before I could stop him. He put it across his knee, and strained it. 'Look,' he said. 'It bends.'

'Give that back!'

A tussle developed; I grappled with him, anger lending me strength, repossessed the sword and sat back panting.

'Can you use it?' he said indifferently. He picked a stem of grass and started nibbling it, spitting the pieces in the general direction of my feet and watching me all the while from the corners of his eyes.

I said, 'Publius, don't do that, please.'

He spat again. 'Publius,' he said mockingly, 'don't do that, please . . .' He sat back on his heels. 'We could fight perhaps,' he said, 'if you weren't a coward, and only armed with a toy.'

My mouth had become dry, as it always did when I was beside myself with rage, and the blood sang so loudly in my ears that the shrilling of insects round about sounded tiny and far-off. 'We will fight,' I said, 'if you can find a toy of your own.'

An odd expression came over his face. He seemed to consider, narrowing his eyes slyly at me; then he smiled. He said, 'Wait here,' and ran back out of sight round the corner of the villa wall.

I waited, half-lying on the grass, plucking tufts of it in my rage and squashing the stems between my fingers. A few minutes later Publius was back. He was lugging a sword, its blade bright with rust. A Legionary sword.

For a moment I thought he wasn't serious; then I saw his face. I got up slowly, my own anger all but stilled; and I think for the first time in my life I sincerely prayed. Not, oddly enough, to the God I had been taught to worship since my earliest days, but to that Shade that once had visited me, the Shade with the gentle voice and cold, shrewd eyes.

We stood facing each other a long time without speaking. Publius swallowed; and I tried to keep my arms from trembling. I couldn't take my eyes off the sword he was holding. I realised, for the first time, that he was going to try to stick that pointed piece of steel inside me, right inside; and I was going to try to do the same to him.

He began stepping sideways, a pace at a time. I tried to turn with him, keeping my sword-tip up in front of my face. I knew now just what Marcus had meant; and that there was no drawing back. I wanted to be sick.

'Now, Sergius,' said Publius, 'you don't deserve it, but I'm going to give you one more chance. I know you're afraid, so just throw your sword down on the grass, and go away. I promise you won't be harmed.'

I swallowed, and said nothing.

'Come on,' said my enemy. 'I shan't think any the worse for

it. I don't expect you to know much about fighting; after all, your father is only an inspector of drains.'

I charged.

My first rush, compounded as it was of fear and desperation, nearly took him by surprise. The swords clashed; the heavier weapon, sliding down my blade, bent the guard, trapping my fingers painfully. Instantly, the rage was back. I circled, heart thudding, watching his eyes. He feinted, stepped back, feinted again and swung the sword two-handed. I caught the blow; and the weakened cross-piece gave way, flying across the grass. The blade edge, red and serried, buried itself thirstily in my hand.

The world seemed to stand quite still. I looked down. I felt no pain; but my little finger, split from the rest, hung uselessly from my palm, and blood was dribbling brightly. I reached across, shaking, to press the wound with my fingers. The blood coursed instantly to my elbow, splashing on the grass at my feet.

My adversary's face was chalk-white, but his eyes were glowing at the sight. 'Fight,' he said, 'little freedman's son.' I raised my weapon somehow, made to run at him; and my sight flickered, fast as the flapping wing of a bird. Earth and grass rose to meet me, and I fell forward into night.

I opened my eyes, expecting to see above me a gulf of sky. Instead there was a pale, plain ceiling; the ceiling of my room.

I tried to move my arm. Pain came at once, increased to a confused burning. I closed my eyes, groped cautiously with my left hand. My right arm lay at my side, the hand swathed in bandages from wrist to fingertips. I tried to lift it. It wouldn't move; it was as if I was paralysed. I cried out and tried to sit up. A spasm of giddiness seized me; I fell back, afraid I might faint again, and lay still a while.

The light was dimmer when I woke, and my mother was in the room. I spoke to her. She ignored me. I spoke again, louder, wondering if my voice had failed as well. She paused at that, glancing up with the oddest look of startled pain before stooping to busy herself with the bedclothes. Something in her expression quelled me, so that I didn't speak any more; and she left the room without answering me. When she had gone I lay quiet, puzzling my sluggish brain. Her face haunted me; the stillness of it, the eyes huge in the lamplight. Like another face, half seen, more than half forgotten. Never before had she refused to come to me; I fretted with the memory, trying to understand.

The room was dark before I thought of the deerhunt. The first tears trickled instantly, salty and hot; then I was sobbing to myself, bitterly and in silence.

There had been a doe at the end who had struggled partially from one of the steep-sided pits. As she scrabbled and fought for purchase a beater, beside himself with excitement, rushed forward, brought a rock down crushingly on her skull. She turned her head to him then, as if questioning mutely why it had been necessary to cause her so much pain; so that his arm faltered for an instant before the stone fell again and she slid quietly back into the earth. I was undoubtedly overwrought; but it seemed my mother's face had been the face of the deer.

A fever came, and ebbed. Braziers were burned in the room, filling it with bitter smoke. Calgaca fed me broth, carefully, propping me upright so I could eat. She changed my dressings, always with her lips compressed into a rigid line. Once, when she turned away, I saw the sheen of tears on her cheek. I lifted my wrist, supporting it with my other hand. I saw my little finger thrust up stiffly now, out of line with the rest; at its base was the curving dark edge of the wound, surrounded by inflammation and areas of wrinkled dead skin. Calgaca applied herbs, and a balm with a stinging scent. I tried to move the finger. I could not; the attempt merely started bright forked pains.

The pain too passed, by degrees. I lay a week or more, in a cold dream of my own. The housepeople brought me my food; nobody else troubled me, or came near. Through the doorway of the room I could see a corner of the peristyle, a wedge of blue sky across which billowed majestic silver clouds. Noises reached me from the rest of the house; the clattering of utensils, voices raised in laughter, argument. My ears recorded the sounds indifferently.

Finally I could stand my solitude no longer. I rose shakily, dressed and went in search of Marcus.

I found him as usual in his room, shaping the handle of a scythe. He stared at me for a moment when I entered then went on patiently scraping, whistling tunelessly between his teeth.

'Marcus,' I said, 'I've got to talk to you.'

He said calmly, 'Nobody's stopping you.'

I opened my mouth, shut it, tried to think. Then it all came with a rush. 'They all hate me. My mother hates me. They're all laughing at me. Because I was a coward. I wish I'd died, I don't know what to do.' I clenched my fists; and to my horror and disgust, my eyes once more started swimming with tears.

Marcus put down the spokeshave he had been using, released the work from the cramp. He said, 'Your mother doesn't hate you. She's merely put you away from her. Which, perhaps, is not before time.' He squinted along the beautiful curve of the haft, nodded, and placed it to one side.

I said dully, 'I wish I were dead. Please help me, Marcus. What am I to do?'

He grunted. 'On the subject of cowardice,' he said, 'you were defeated by superior force. Not particularly smart, maybe; but cowardly? I wouldn't have said so.'

I said, 'I lost my sword. All because of the blood.'

He stared at me, and sighed. Then he turned away, drawing his key ring from his finger. Above the bench, built into the corner of the little room, was a stout cupboard of unpainted wood. I had never seen it open. He unlocked it, swung the door ajar; and I caught my breath. The light gleamed on leather and steel; on the trappings of a soldier, of the army of everlasting Rome.

He took the heavy sword down in its scabbard, pulled the blade clear of the sheath. I flinched at the sound it made. He turned with it in his hand; and this was no toy, no old blade dull with rust, but a live thing, bright and polished and terribly keen. He held it upright, the hilt between his palms, and thought for a moment. 'Answer me a riddle,' he said. 'When a great fire burns, in the forest, what do people do?'

I said, 'I don't know.' My new fancy toy of rhetoric had completely failed me.

'They light other fires,' said Marcus. 'And these lesser fires burn back towards the greater till both are spent. Do you understand?'

I shook my head, tiredly.

'Fear is like a fire,' said Marcus. 'Something bright and crackling, that can burn a man to a husk. I've felt it, times enough; but never so sharply, and never so hot, as the day I first saw my own blood running on the ground.'

I looked up sharply, wondering. The notion that Marcus

could ever have been afraid of anything had never occurred to me; it seemed impossible.

'I made a vow that day,' he said. 'I swore that whatever happened, I would shed no more. And I found a lesser fire to burn away the great fire in my brain.' He took my wrist, drew it towards him and closed my fingers on the sword-hilt. 'Touch it,' he said. 'Hold it firmly now. Feel its weight. How it balances. Never be afraid of a sword again.'

He left me then, came back with a tray on which stood bread, a bowl of soup and a flask of his famous tart wine. He said, 'You did well to come to me. I was hoping you would. Tomorrow I'll start you on some decent training. Now wipe your nose, and get some food into you. If I catch you snivelling again I'm going to kick your backside round the entire circuit of the town walls of Italica.'

It was only that night, lying in the dark remembering what he had said, I realised that though the things he had warned me of had come to pass as he expected, there had been no word of blame. I don't think I ever loved anybody as much as I loved Marcus then.

And now I must span years, in which my life and schooling in Italica went on; rich years, I see now, while I grew from childhood to my first strength. Under Marcus I learned a soldier isn't made in a few days, or even months. I had dreamed myself a Caesar, dreamed the spoils and glories; what I hadn't dreamed was the sweat, the tedium, the endless repetition and disappointment of the training he gave me. Most days, sometime or other, we would ride out of Italica to our practice ground, a glade in the spinney in which I'd played that fateful morning. My targets were stakes hammered into the ground; he would keep me in play an hour or more, till my muscles sang and my whole body ached. He gave his praise grudgingly; I learned to put my heart and soul into the work, and every last ounce of my strength. 'When the steel's in your hand,' he would say time and again, 'it's no good looking for quarter. Only one thing matters; to stay alive. . . .'

His first concern was for my grip, for the injured finger never regained its use, and any backhand stroke tended to flip the sword up out of my hand. He devised a special hilt for me finally, angled slightly to the blade and with a curved tailpiece that fitted snugly in my palm. He had other tricks, too many to recall. Once he asked me, casually, whether I felt my strength had improved. I answered smugly enough that I could fight for half a day and not feel tired. He gave one of those humourless little smiles of his, and said no more; for the next practice he tied slabs of lead to my shield and blade, and I found just what it cost to make a thoughtless boast. Another time I confided to him that were I to meet my enemy again I felt sure I could beat him. He said nothing in reply, but I should have been warned by the expression on his face. Next time we fought he changed his tactics, barging me violently and hooking a foot behind my ankle. I landed sprawling; before I could roll aside the tip of his sword had kissed me lightly on belly and throat. 'It's not the man that's worse than you that you must look to,' he said savagely. 'It's the man that's better. . . .' After that it was practice with the stakes again; he wouldn't deign to fight me till I had once more been reduced to a true appreciation of my worth.

Marcus always brought to practice a bottle of his sour wine; we would share it when training was through, sitting side by side, while he sketched points of tactics with the tip of his sword, dredged up some anecdote from his seemingly endless supply to illustrate a point he'd made. Other times he would discuss the styles of fighting of barbarians, frequently as we worked. 'Your Libyan,' he would say, 'won't stand unless he's three or four to one. Don't let that bother you. He'll come at you bollock-naked anyway, not even a shield; just a bit of cloth wrapped round his arm. He carries one light spear; when he's cast he'll turn and run. A German's a very different proposition. He'll barge you, try and get you off your feet; and that bloody little sacred dagger he always carries'll be in your gut before you've got time to blink. Come into him hard, shoulder and shield; and use the point. Always the point. Shield up, keep it up, and thrust. Into me now. The point, and again. Give me the point. . . .'

Sometimes I worked with the trident and javelin to improve my balance and eye, but always we returned to the sword. Hour after hour, day after day; till one day I disarmed my instructor, sending his blade spinning to the ground. He straightened then, and laughed loud and long. When he had finished he shook his head. 'There'll be no more lessons,

Caius,' he said. 'Only practice. I've taught you all I can. Except one thing.' He drove his shield edge suddenly and violently at my stomach. I sat down with a thump, winded; when I looked up the tip of my own sword was pointed unwaveringly between my eyes.

'That's my final warning to you,' said Marcus gruffly. 'Never trust an unarmed man. . . .'

It was only afterwards, riding back to Italica in the mellow sunlight of a summer evening, that the true significance of what he had said dawned on me. I glanced across at him, sitting his horse easily at my side; but his brown, keen face was remote and composed. He had given himself unstintingly; now his task was through. And it seemed another phase of life had ended for me before I could realise its passing. For years the daily practising had been an end in itself, something I had grown to do instinctively and automatically; now, abruptly, the future yawned blank as that first day at school. I was seventeen, tall for a Roman, strong well beyond my years; and the time had come to put what I had learned to a use. I remembered something Marcus had said once, about sand never running upwards; how both men and nations must go on, looking to the future and never standing still.

The time was coming when I was to wish with all my heart I could conjure the sun back up from the western sea; for my mother was pregnant again.

Chapter three

While I worked to fit myself for my chosen career, great changes were taking place in the West. I was thirteen when Magnus Maximus, for so long a thorn in the side of the Imperial Government, finally rode to meet his fate, invading Italia with a horde of foederate barbarians stiffened by regular troops drawn from Hispania, Britannia and Gaul. Defeated in pitched battle by Theodosius, he finally surrendered at Aquileia and was summarily executed. For a time the Empire breathed easier.

Yet the position of Theodosius himself was still delicate. Latium and many of the Provinces recognised the legal rights of the House of Velentinian; the Augusta Velentina, ruling for her son, was still a voice in world affairs, while her Magister Militum, Arbogast, was both powerful and feared. Theodosius, whose weakness had never been better shown than by his initial recognition of Maximus as Augustus, took the opportunity presented by the crisis to strengthen his position in the West. For three seasons his vast field army policed Gaul; but early in the summer of my seventeenth year news reached Baetica that the Legions were once more on the move. The Emperor was retiring on Constantinopolis, which city his son Arcadius had been controlling in his absence.

My father put his head in his hands when he heard, and swore with startling fluency. 'It's incredible,' he said. 'Absolutely bloody incredible.'

A weed was growing between two of the flagstones; I kicked at it idly, swinging my feet.

'What does it mean?' I asked.

'What does it mean?' said my father. 'It means just this. Theodosius had the chance to clear things up for good and he let it slip through his fingers. Arbogast is a barbarian. He can't be trusted; and that the Augustus knows as well as anybody.' He swore again. 'He should have destroyed him,' he snarled, totally forgetting his vaunted liberality. 'And if necessary, young Velentinian and his mother as well. This is what comes of too much religion.'

I said, 'What's religion got to do with it?'

He favoured me with a steely glare. 'What I shall never understand,' he said, 'is why this education of yours has failed to instil the slightest awareness of what's going on in the world about you. Theodosius was taken seriously ill, years ago; he thought he was going to die, and had himself baptised. Since then he's surrounded himself with an entire corps of Bishops, all prating on Heaven, Hell and this sin, that and the other; and it's them he listens to rather than his military advisers or his own common sense. We haven't heard the last of that clan of Pannonian misfits, not by a long chalk. You mark my words.'

His face altered.

'But I forget myself,' he said viciously. 'Politics have always been a little too mundane for your taste, haven't they?'

I stared at him, finding no answer. I saw the balding head, the lined face, the cold, angry eyes; and I realised I had grown farther from him with every year that had passed.

I got up and wandered off to find Calgaca.

Over the years my relationship with my mother had undergone a subtle change. After the initial shock of my injury, when I had been carried home unconscious and smothered with blood, it seemed a deeper bond had grown between us. True, the old days were gone for good; I no longer expected Calgaca to come running if I cried out in the night, those things were ended. But instead, as she recovered from the strange grief that had oppressed her, she began to treat me more as an equal. I would sit with her by the hour, often at night after the lamps had been lit, chattering about all manner of things. I would report to her, in detail, on my training; and though it frequently pained her to hear of my growing skill she was always careful to show an interest, and quick with questions and praise. She could never hear enough about the day-to-day events of school; in fact I would probably have tried to secure an early revenge on Publius had it not been for a promise she extracted from me. In return she told me more of the history of her side of the family.

I knew that her parents had originally emigrated from Britannia, but I was surprised to find what a hold the place still had on her affections. She sometimes described it to me, sitting with her hands quiet in her lap, eyes watching thoughtfully out across the darkening peristyle. 'You'll hear often enough that the climate is very bad,' she would say. 'It isn't. It's beautiful. The skies are veiled and misty, not hard and brilliant like the sky over Italica; and the grass is richer and greener than any you see here. Evenings and mornings are often the loveliest part of the day. Especially evenings. The sun doesn't rush into the sea like it does here in the south, as if the Devil were behind it. You have time to watch the shadows lengthen, and the whole land slips into the dark like a countryside seen in a dream.'

How much she truly remembered, I couldn't say; for she was very tiny when she left, not much more than four or five years old. In Britannia, too, her family had been breeders of fine horses; but the persistent raiding of Picti from the north and pirates from the east, from what they call the Saxon Shore, had rendered the raising of livestock too hazardous to be profitable. Julian eventually sent an expeditionary force to dispose of the nuisance, but by then my grandparents had had enough. They owned considerable wealth, mainly in the form of gold and silver plate; the heirlooms, melted down, secured passage to Gaul for them and the best of their stud. Some of the horses died in the crossing, but enough survived to re-establish the line in Hispania. The new blood proved highly successful; and had it not been for the generally impoverished state of the Empire the family would rapidly have restored their former fortunes. 'As it is,' Calgaca would say, 'it sometimes seems we exchanged one set of troubles for another just as bad. That's why I'm grateful to your father for providing us with such a fine living and home, and why you must be too.'

I remember one evening spent with her particularly well. It was a hot summer night, shrill with the chirring of insects. She had had a chair carried out to the peristyle, where the air was a little cooler. She was working on a new altar cloth for the local Bishop; embroidery was one of her favourite occupations.

Her fingers moved deftly, busy with the needles and coloured thread; I sat at her feet, watching the last of the afterglow fade from the sky. The lamps, burning steadily, cast gently moving pools of light on the columns and pale walks of the place; above the shield walls of the little court, the blue of dusk seemed intensified by contrast. That was the night she first told me of Tir-nan-Og, the Land of the Blest. 'When I was very small,' she said, 'I used to stare out over the sea, usually about this time, and imagine I could see it far away in the west. They used to tell me it was a cloud bank that came at night, low down on the water, but I always knew better. In the morning it would be gone; but it always came back. All the peaks and valleys of it, shining in the sunset like gold.'

I frowned. 'Mother,' I said, 'I don't understand. What did you say, Tir-nan-Og?' There was silence between us for a moment; then Calgaca laughed, and leaned forward to smoothe my hair. 'Don't tell your father I talk to you about such things,' she said. 'I'm afraid he wouldn't approve.'

I waited again, for her eyes had darkened, assuming the misty, far-off expression I had come to know so well.

'Tir-nan-Og,' she said finally, 'is the oldest faith of our people. And we are a very old race, older even than Rome. Years before she was ever built we were worshipping Gods of our own; some of them very like the Christos we're taught to pray to now. We had a Heaven, too. It lies over the western sea, a very long way away; so far that no boat has ever sailed to it, or ever will. For us, it is the Land of Heart's Desire. There you can find eternal sunshine, eternal freedom and peace; and the lucky ones who reach it never grow old.'

As she spoke she had slipped unconsciously into the present tense; so I saw the legend was still a living thing to her, and very close to her heart. I sat still, unwilling to interrupt; but when she paused again I stirred myself. 'But, Mother,' I said, 'if no boat can sail to your land, how does anybody ever get to it?'

She waited, still with the strange, withdrawn expression on her face. Behind her the image of the Virgin watched from her plinth. The lamp flames wavered in a stirring of breeze; as the light played and shifted over the statue its expression seemed to alter. It was as if the stone itself had come to life, and was listening.

My mother laughed softly. 'There is a Boat,' she said. 'A

white Boat, with no oars, that needs no wind to drive it. But no man can order its coming.'

Her hands were still now in her lap. 'I dreamed of it once,' she said. 'It was sunset, all the water glowing for miles and miles like a new, polished shield. There seemed to be a great crowd of people on the shore. And one by one they stopped and stared, and pointed at the sea. And there it was in the distance, so small at first I could barely make it out. It was like a swan, and golden in the light. There was no wind to send it in. Even the waves on the shore fell softly, like ripples in a brook.'

She was quiet again for a long time. I turned my head to look up at her face. Her eyes had widened with the intensity of what she saw; she seemed unaware either of my presence or the house in which we sat. Suddenly, I was uneasy. A shiver ran through me, so that I spoke quickly to break the silence that had fallen. 'Why did it come, Mother?' I asked her. 'The Boat? Was there any meaning in the dream?'

She nodded slowly, lips half parted, not looking down. 'Yes,' she said, 'there was a meaning. My sister's soul was on the machair. Only then I didn't know. In the morning the Boat had gone. And I never saw the Land again.'

She shook herself then, as if from a heavy spell. 'Well, Caius,' she said, 'I haven't the slightest doubt your father would call all this a load of nonsense. And who's to say he isn't right? Run along to the kitchen, will you, please, and ask Ursula to bring us two cups of wine. Then I think I'll go to bed. I'm feeling a little tired.'

I spoke impulsively. 'Mother,' I said, 'one day I'll take you to Britannia. I promise. I'll promise by the Virgin, if you like.'

She turned to me, smiling, and reached to touch my arm. 'Never make vows, Caius,' she said, 'unless you know you can keep them.'

I answered a little awkwardly, unsure of my ground. 'I don't, Mother,' I said. 'At least, not to you.'

I glanced at the Virgin again as I passed. For the first time she too seemed to be smiling.

Calgaca's pregnancy came as a most unwelcome surprise. I had assumed for some reason that she was past the age of child-bearing; and in any case the doctor and midwife who had attended her at my own birth had warned her it would be dangerous for her to conceive again. Some deep part in her tore as she voided me, so that she nearly died, and for months afterwards could neither walk nor stoop without great pain. Now I blamed my father, bitterly, for the danger in which he had so casually placed her; but when I mentioned this to her she frowned and gripped my arm. She was resting on her bed at the time; she looked up at me sternly and said, 'Caius, for Heaven's sake don't be so silly. It's perfectly right and natural for your father to want another child. In fact we both did, and I for one was pleased when I found out I was expecting again. You should be as well. You're going to have a sister; and she'll grow up to be very beautiful, I can tell already.' But even as she spoke a spasm of pain gripped her; I saw her pale suddenly, and bite her lip. 'You'd better go now,' she said, 'I want to have a sleep. But before that, will you ask Ursula to come in? If you want to make yourself useful, and please me as well, you can help Marcus. I asked him to clear one of the spare bedrooms out for the baby. Go on now, Caius; and whatever you do, don't go about with a long face like that. You'll only get your father annoyed. When you know more about these things you'll realise how easy it all is.'

But for my mother carrying the child was far from the simple affair she made out. As the year wore on, and she came nearer her time, I would lie awake at nights listening to her moaning in her sleep. Once she called out loudly, bringing the housepeople scurrying to her room. They wouldn't let me in; I crouched outside, cursing and praying by turns and starting at the slightest sounds. An hour or more passed before I was told the crisis was over. I had hoped, sincerely, that Calgaca would lose the child; I was beginning to hate the unborn baby too. But it was not to be.

My father remained his usual self; courteous to his associates and the servants, cold and distant with me. I avoided him as much as possible, spending all the time I could at my mother's side. She talked continuously now, when the griping pains would let her; about Britannia, and her childhood, and the strange land across the sea. Sometimes it seemed her mind wandered, so that she would burst into that other tongue, the lisping chatter my father had forbidden her to use. Then she would roll her head and groan, gripping the bedcovers with her fists; and Ursula or one of the others would come bustling, shoo me out in spite of my protests. At such times I raged at my helplessness; when my fear became too much to bear I would take a horse, ride like a demon to our old practice ground, hew and slash for an hour or more at a stake till the sweat ran from me and my body shook with fatigue. Sometimes Marcus would come with me; but usually he left me alone. Nobody could help me, any more than I could help Calgaca.

A doctor was sent for from Gades to assist at my mother's delivery. Typically, I knew nothing of it till the raeda came clattering up to the door. I suppose he was a good man in his way; he dosed Calgaca with drugs, extracts of poppy and laserpicium, and advised complete rest and quiet. But he denied her my company; and I hated him for it too.

Things stood at this unsatisfactory pass for some days. Then one morning, just before first light, my mother sent for me. I was up and dressed; I hurried to her room, and was shocked at the change in her. Her eyes, once bright, were lustreless, and pain had drawn deep shadows beneath them across her cheeks. Her hair lay uncombed on the pillows, draggled with sweat; and her hands, resting loosely at her sides, were so thin and white as to seem almost transparent. She must have seen the expression on my face, for she laughed and tried to raise herself. 'Well, Caius,' she said, 'I'm sure you'll be glad to hear all our waiting is over. Your new sister will arrive today.'

The roof of my mouth had dried, so that I had to swallow and wet my lips before I could speak. 'How do you know, Mother?' I said. 'How can you tell?'

She said simply, 'I can tell. Now, Caius, bring that chair over and sit with me for a moment. I want you to listen very carefully. I have an important job for you.'

I did as I was told. When I was seated she lay back, again with a little smile. 'You must have heard from the servants' talk,' she said, 'that women get odd fancies at times like this, and like being pandered to. So will you do me a favour?'

'Of course,' I said. 'Anything, Mother, you know that. Anything at all.'

'Good,' she said. 'Then I want you to go and find Marcus. He knows what it's about, I've already spoken to him. I want you to go hunting.'

For the moment I was too surprised to answer. It was about the last thing I had expected to hear. Eventually I said, 'But where, Mother? Where?'

'To the north,' she said. 'Where you always go. I want you to shoot me a deer.'

My jaw must have sagged. Nothing I could think of would have taken me from her side at that time. I started to protest, but she stopped me, raising her hand. 'Caius,' she said, 'a moment ago you were promising to do anything I asked. Are you going to go back on your word already?'

I couldn't answer. I shook my head, dumbly; and suddenly the tears stood in my eyes, spangling my vision so that all I could see of her seemed to shimmer and dance. She reached out then, gripping my wrist, and for the first time let her fingers wander down across the curving scar. She watched me steadily, lips still quirked into a smile. 'A deer, Caius,' she said. 'The tenderest you can find.'

A lump had risen in my throat, so that I had difficulty in speaking. I said eventually, 'Will you be all right, Mother? Will you promise?'

She let her eyes drift shut for a moment, opened them again. They searched my face, moving in puzzling little shifts and changes of direction; then she held out her arms. I embraced her, gently, till she pushed me away. She watched me again; her face seemed altered now, younger and less strained. She said, 'I promise. When you get back it will be over, and I shall be sleeping. Tomorrow I shall want my venison. Go now, dear, you have a long way to ride. Go carefully.'

I stood looking down at her for a moment. It was as if her image was burned in some strange way on my brain; for I can see her still. Then I turned away, pushing aside the drapes that closed the room. I couldn't trust myself to speak again.

I found Marcus was ahead of me. His room was empty; as I ran back to the atrium I heard the sound of hooves in the street. I changed hastily into a short riding tunic and high, stout boots, collected a bow and my sword and left the house, calling to Ursula as she hobbled by to look after my mother well. Outside, Marcus sat grinning broadly. He was leading my favourite mare, a big-boned grey from my uncle's stud at Augusta, and a packhorse. At its heels skirmished a brace of Egyptian hounds, lean yellow creatures elegantly feathered on tails and ears. I mounted, clattering past the front of the house. Septimius, already cudgelling his reluctant staff into action, waved to me cheerfully;

I returned the salute, setting off at a fast trot for the west gate and the Lusitania road.

The sun was still barely up. The air struck chill; but the haze in the sky promised a scorching day. We rode steadily, taking advantage of the coolness; by mid-morning we were well clear of Italica. We turned north from the road, crossing sweeping uplands to the wooded country that was our objective. My spirits had revived; so that when we stopped, in the shade of a massive outcrop of rock, and Marcus produced his inevitable skin of wine, I could laugh and joke with him as of old.

After the brief halt we hoisted the dogs across the fronts of our saddles, to conserve their energy. Marcus had trained them to lie like that; it was a trick he had picked up in the east. We rode on again, and by midday had entered the outskirts of the forest. The trees arched above us, spreading a leafy shade that was welcome after the torrid heat of the plains. We dropped to a walk, watching for game paths, engrossed in the delicate business of sighting and stalking our quarry.

The woods teemed with life. We saw several herds of the little dappled deer, but each time they took alarm before we could come within bowshot and Marcus refused to loose the dogs. It was early evening before we killed, in a little clearing deep in the forest. Marcus saw the herd first, their spotted coats blending perfectly with the moving patterns of leaves and sunlight. He raised a hand, warningly; but the next moment the lead buck had lifted his head and snorted, and was leaping off through the trees. I rode hard across his line, yelling to the dogs; for the Egyptian hound hunts by sight. Fortunately our pair knew their business. They were away in a flash, running silently, heads low. We were about to follow them when a doe, separated from the rest, broke back blindly across our path. There was no real time to aim; I loosed at her hardly expecting my shaft to strike home. The arrow took her high on the shoulder, piercing her to the heart so that she fell with scarcely a kick. I was delighted with the shot; Marcus dismounted and ran to her, knife ready to slit her throat. Despite all my training, I had never quite brought myself to do such a thing. In the meantime the dogs had brought a fine young buck to bay. We hurried to them, guided by their high-pitched yaps and squeals. A second shaft despatched the frightened creature cleanly; Marcus bled and butchered it on the spot, wrapping the best cuts of meat first in fresh leaves, then in the hide. We returned to the clearing with the excited dogs, and built a fire; we had carried live coals all day, in a plugged earthenware tube.

The little glade was golden now with sunlight, the beams reaching mistily between the trunks of trees. I lay back on a sloping bank of grass, my shoulders resting against the mossy roots of an oak. As the smell of roasting meat mingled with the scent of the fire my mouth began to water uncontrollably. I realised I was ravenous.

Then, quite suddenly, an inexplicable thing happened. Nothing changed, in the clearing; but it seemed a chilling wind reached me across the grass. Leaves rustled on the massive branches above my head; it was as if a voice wailed, wordless and in pain. So strong was the illusion that I sat up, appalled. Marcus, squatting turning a hunk of meat above the fire, looked round startled. 'Sergius,' he said, 'what the Devil's the matter?'

I couldn't answer. The sun still shone; but silvered now, and lacking warmth. Above me the trees seemed suddenly disfigured and monstrous, thrusting vast claws at the sky. The wind came again, moaned away over the forest-tops like a thing lost and afraid, passed into distance, and was gone.

I groaned aloud, and ran for my horse. 'Marcus,' I said, 'what fools we've been. Oh, fools . . .' I heard him shout behind me. I ignored him. I was already riding, driving my heels at the creature's sides, beating its neck with my fist. Branches swooped at me, lashing my face; I ducked beneath them, flinging them aside with my free hand. The sun, low now, dazzled me, flashing between the trunks and leaves; I gave the animal its head, urging it blindly south towards Italica.

How I escaped breaking my neck in that wild ride I shall never know. The blood pounded in my ears, deafening me to reason; it seemed my fear communicated itself to my mount, lending it wings. I was clear of the forest before Marcus overtook me; I would undoubtedly have killed the horse had he not ridden alongside, lunged dangerously to grasp the rein. The animal bucked and plunged, nearly throwing me, slowed at last to stand trembling and tossing its head.

Marcus was furious, quite justifiably, having lost the meal he had cooked so patiently and risked his life into the bargain, in the crazy charge through the woods. 'What the Hell's got into you?' he shouted. 'Have you gone completely off your head?' I answered, panting, that something was terribly wrong, that I must return to Italica at once. He tried to talk sense into me, of course, calling me every sort of bloody fool; but I must have looked so queer that finally he gave up arguing. He whistled to the dogs and let me go, trotting grimly a yard or so behind.

We made what speed we could, while the sun dropped to the horizon, the long shadows raced forward across the plain. The ride, that in the morning had passed so pleasantly, now seemed endless. The horses were tired, of course, after the long day's work; it was full night before we struck the Lusitania road, turned east towards Italica. When the town walls finally came into sight I could no longer restrain myself. I forced my mount to a canter, clattered through the gateway and along the street. Passers-by cursed me, tumbling out of the way; I ignored them, reining to a violent halt outside my home. The horse was lathered; I flung myself from the saddle, let the animal run free. I heard Marcus swear as he rode after it. Shops and porch were in darkness, but beyond a wavering reflection told me the house glowed with light. I ran for the servants' door.

As many lamps as we possessed burned in the atrium and the rooms leading from it; beyond, in the peristyle, torches cast a flickering orange glare. There was a scurrying of feet, voices calling; and closer at hand, rising and falling monotonously, the high-pitched noise of weeping.

I nearly collided with Ursula. I grabbed for her arm; but she backed off at sight of me, hands clapped to her throat. I ran then for my mother's room; behind me, the sobbing redoubled.

There were more lamps, burning in niches to either side of the door. As I reached it a woman backed through, arms full of a great bundle of linen. She turned, stricken, at the sound of my footsteps; and the lamplight shone full on her burden. The whole mass gleamed, wet and terrible; and stank, and was red.

They had drawn a cover over her, across that fearsome bed. Her eyes were closed; and it seemed the lines and tiredness had been smoothed from her face, so that as she had promised she was merely sleeping. I lifted her in my arms. She was warm; her head rested heavily against me. I stroked her hair, calling to her. It seemed absurd; it seemed her spirit, hovering so close, must hear, force its way back into the flesh. I called again and again, till the little room rang with the noise; and her head fell back, showed me beneath the lowered lashes the glinting lines of white. I knew then, whatever summoned me in the wood, her soul had never found her western land. It drowned in the blood the poor clay dropped, died unwanted and alone.

Someone finally dragged me away from her. I don't know who. Marcus, probably; nobody else would have dared touch me. Of what happened in the dreadful hours that followed I have no clear recollection. For me, the house was haunted. Doors swung with no hand to guide them, drapes stirred to non-existent winds; while I called Calgaca, over and again, hoarsely, beating my fists bloody on the ground. I heard the mumbling priests at her door, the high-toned useless prayers, the shrill wailing of the housefolk. In time the place was quiet. I rose from where I had flung myself across my bed and went to find my father.

Lights were burning in his study. I opened the door unannounced, closed it behind me. He sat at his desk, reading. Reading, from a book, while my mother lay still warm. He looked up as I entered, closed the thing gently and laid it to one side. He began to speak, but I cut him short.

Again I have no real memory of what I said. It seems the word 'murderer' passed my lips; if it did, it was without my volition or control. His reaction was swift. I was leaning over him, shouting; he half rose, caught me a blow across the face that knocked me off balance. I sprawled against the wall, straightened slowly feeling the burning in my cheek and jaw.

He too was on his feet, but his eyes weren't on my face. I looked down, following the direction of his glance. My dagger, the little dagger I always carried, was in my hand. I stared at it, bemused; to this day, I swear, I have no recollection of drawing it. I suppose I could have killed him, in that same mad flush, and brought away scarcely a memory.

There was a long silence; then he said quietly, 'I see . . .' He seated himself again, slowly, spread his palms flat on the desktop; and it was as if we acted out some grim and fabulous play, all the lines learned wearily by heart. His words fell measured and steady, like little stones; before his lips had shaped them I knew what they would be.

'Caius,' he said, 'many years ago I instructed my servant to speak to you. It was also my wish that he train you, in swordplay and the general use of weapons. My hope was that it might make a man of you, but it seems the experiment has failed.

'For too long now you have deliberately flouted my wishes. I let you go your own way, hoping that one day you would come to your senses. In that, too, I have been disappointed. Now you have shown me steel, under my own roof. You have gone too far.'

He drew a heavy breath before continuing; and already the fear was on me. Fear, and a burning self-contempt. The desolation I felt was not for Calgaca but myself; so soon had selfishness overcome my grief.

'You will leave this house,' said my father. 'And your instructor, who has taught you some things too well, can go with you. You will pack whatever you wish to take at once; I want you to be gone by dawn. If you remain, I shall take steps to secure your removal. This is no longer your home.'

The shock, coming so soon on the greater shock of Calgaca's death, almost unmanned me; a part of me wanted to fling myself at his feet, grovel and beg forgiveness. Yet so swift is thought that I understood in the same instant the uselessness of such a course. With the realisation came a cold, swimming rage. It dictated that I should use the weapon I had drawn, add my father's blood as a sacrifice to the blood already spilled. I think he saw that; none the less, he droned on.

'I shall not disinherit you,' he said. 'If and when you acquire maturity, and some sense of duty, you may return and ask my pardon; until then, I don't want to set eyes on you. I shall give you no money; you must be prepared to make your own way in the world. However, you may if you wish take a horse from the stables; I shall expect you eventually to return its price to me. Where you go is entirely your concern; but I will if you choose give you a letter to my brother in Rome. I shall not recommend you to him; he must form his own impression as to your usefulness. And may God help you. Now you can go. And send Marcus in to me, please.'

He looked away at that, picked up the book and started to read. He never glanced up again; and I turned and left, without a word.

Marcus was up, sitting brooding over a jug of wine. He

glanced at me as I entered; then rose, silently, at the expression on my face. I told him, flatly, what had happened. I think he was as shaken as I was; when I had finished he stood for a moment, pursing his lips and frowning, before reaching silently to grip my arm.

'Sit down,' he said. 'Wait for me here. I shan't be long.' He strode out, quickly. I heard his feet on the flags of the atrium, far-off the click of my father's door. Then there was quiet.

I sat listlessly, watching the lamp flames bob and dip. In time, he returned. He didn't speak to me; just crossed to the workbench and began quietly stowing the tools, setting chisels, hammers and saws neatly in their racks. When he had done, and the wood shavings were swept into a pile, he lifted down two heavy saddle panniers from the wall. He moved round the room, quickly and efficiently, selecting weapons and clothes, stowing them with the ease born of long practice. I watched him uncomprehendingly for a time before I spoke. When I did open my mouth my own voice was a shock to me. 'Marcus,' I said, 'we must go to Rome.'

He stopped at that, tight-lipped, and looked down at me. His face in the lamplight was grim and hard. Finally he shook his head. 'No, lad,' he said, 'that's where you're wrong. We don't have to go anywhere. You can take yourself to Rome, if that's your wish; but I don't have to trail along after you. I was your father's man. I served him well enough, to my way of thinking; and precious little thanks I've had for it, when you weigh it through. Well, that's one thing; I suppose I should be used to ingratitude by now. But I'm not thinking of taking service again; least of all with you.' And he wrenched up the lid of the strong-box, began stowing coins into the pouches of a body belt.

I nodded, realising that God, or the Gods, had not yet done with my punishment. I drew my dagger again, sat looking dumbly at the blade. The old fear of shedding blood was on me strongly, but I think at that moment I could readily have opened my veins. 'Then I shall join my mother,' I said. 'I'm sorry, Marcus, for speaking stupidly.'

He turned from what he was doing, thoughtfully; then reached across and plucked the weapon from my hand. 'Go and pack,' he said gruffly. 'And let's have no more talk like that. If you want to please your mother, you can start behaving like a man;

it's not too late.' He paused, momentarily; then his voice softened. 'Rome's as good a place as the next,' he said. 'And if we're travelling the same way we might as well ride together. Your father made me responsible for you years enough ago; I know I'm a bloody fool but I don't like leaving a job half done. Go on, get on with it. Come back when you're through.' And he readdressed himself to the stowing of the coins.

I walked dazedly to my room. But the effort of logical thought was too much for me. When Marcus put his head round the door an hour or more later, I was still sitting on the bed, surrounded by the litter I had pulled arbitrarily from cupboards and chests. He set to without a word, making up essentials into bundles, discarding the rest. When he had finished he dumped the packs on the bed. He had laid out a thick tunic and cloak; I changed into them, hearing his voice in the peristyle.

'I wish,' he said quietly, 'to pay my respects to the Domina....'

There was silence, lamplit and flickering, till he returned. I followed him then, hefting the packs. Moments later, we were in the street; the door clicked softly behind us.

He roused Victor at the stables. The groom yawned and grumbled, demanding my father's authority. Marcus ignored him, picking out the two best horses and saddling them silently. He slung the packs across the back of a third animal. 'That's a bonus for good service,' he said grimly. 'I'm paying it to myself.' A half-hour later we were riding through the quiet streets of Italica.

The moon was high, sailing a serene sky. The air struck chill; I muffled myself in my cloak, shivering, following Marcus dumbly and automatically. We passed streets and buildings I had known from earliest childhood; the town baths, the little library, the dyeshop above which Gellius still presided sourly over his classes. The moonlight lay bright in the streets, but nothing stirred. Houses jerked past, their windows blind and dark as the eyes of skulls; it seemed to my fevered imagination the place was already a town of the dead. We passed through the east gate unchallenged, emerged from shadows on to the white, paved road. There we both reined, looking back. Behind us lay my mother and everything I had known; in front the road stretched between tall cypresses, a dim, straight ribbon vanishing into the dark. Somewhere an owl called, haunting and shrill; I shivered again, involuntarily, at the omen.

Marcus clicked to his horse, urging it gently forward. I followed him at a walk, passed into the shadows of the first trees. The hooves rang hollowly on the metalling. Ahead was the Way of Hercules; at its end, seventeen hundred miles away, was Rome.

Chapter four

Dawn found us still on the road. The sun rose directly ahead, flinging mile-long shadows down the paving at our backs. Away to our left loomed the mountains that ring Baetica; nearer at hand stretched the river that gave the district its name. It flowed silently, its broad surface washed with pink. Its banks were lined with clumps of reeds; flocks of duck and bustard erupted from them as we approached, with a clamour of wings. As the sky brightened the flanks and high slopes of the hills seemed to glitter; they hung stark and detailed in the clear air, looking almost close enough to touch.

The sun had lifted clear of the horizon when we came in sight of a cluster of low huts. Marcus rode to the nearest, leaned from his horse to rap at the crudely fashioned door. There was a long wait; he hammered again before the door was opened a grudging few inches. A halting exchange ensued with whoever was inside; then he swung from the horse, gesturing to me to join him. I stooped on hands and knees, crawled after him through the low portal.

Inside, the place was almost pitch dark; the air seemed chokingly thick after the cool freshness of dawn. A fire burned in the centre of the hut; smoke from it swirled in the confined space, stinging my eyes. Some light filtered through the outlet in the conical roof; by its aid I made out rough beds, mere bracken-filled niches in the thick mud walls, a crude table built of slabs of stone. In one of the beds sprawled three or four naked children. Their bodies showed dimly, pale and smooth as the undersides of slugs.

Suspended above the fire was a smoke-blackened pot, from

which the peasant ladled soup into two platters. Marcus passed one to me. I had no desire to eat; I took it from him anyway, not wanting to anger him again. The stuff was hot and thick. Marcus flung down some small coin, and helped himself to another cup.

I had never been inside such a place before. I sat uneasily till he had finished, crawled ahead of him back to the open air. As I straightened up a sharp irritation made me slap at my wrist. I looked down. We had only been inside the hut a matter of minutes, but already I was alive with fleas.

We rested through the heat of the day, pushed on again in the evening, still with the mountains marching to our left. We slept in the open that night; or rather Marcus slept. I lay huddled in my cloak, staring up hour by hour into the vault of sky. I was glad when dawn came and I could rise and saddle my horse. In this way we reached Corduba, two days' travel to the east. Marcus bought a tent there and certain essential stores. In the morning we moved on, following the road as it plunged down to Carthago Nova and the coast.

I was a poor enough companion without a doubt. I rode always a pace or so behind, wrapped in a sad cloud of thought. I relived the events of that terrible day, time after time; my mind, circling uselessly, balked over and again at the monstrous fact of Calgaca's death. I felt her weight against my arms, saw her head loll, smelled the stench of fear and blood that had thickened the air of the room. I heard my father's clipped voice repeat that speech he must have planned a score of times, felt the shame and terror that had filled me. Sometimes I would indulge in fantastic dreams. I would make my fortune in Rome, return rich and powerful, dispossess my father as he had dispossessed me, build for Calgaca the most splendid tomb in the world. At others I saw more clearly how my callowness and stupidity had cost me everything I thought I owned; then I would writhe in futile self-contempt.

Some nights we camped, pitching the tent a little way from the road; others we spent in the inns that bordered it, ramshackle and dirty for the most part and full of itinerant tradesmen, tinkers and the like. The beds, when they existed, were invariably thick with fleas and bugs. I had never been used to such conditions. I was soon covered from head to feet in bites; my face swelled, puffing my eyes half shut, while the tender parts round my crotch became so inflamed that it was agony merely to sit my horse. Even tieing my tunic caused me pain; when I stripped at night the skin round my waist, where the cloth had chafed me, was crusted with dried blood. I bore the discomfort indifferently, as I bore the heat and dust, the glare of sunlight from the endless white road. For me, the world had almost ceased to exist. All that was real lay behind me in Italica; I moved sightlessly, like an automaton, towards whatever fate the amused Gods had next in store.

In Barcino, Marcus managed to find a tavern a little better than those to which we had grown accustomed. We sat there, in the cool of the evening, in a little room open to the street where the owner brought us wine.

I had drunk with Marcus often enough before, but always in strict moderation. That night he plied me with a heavy, unwatered local brew, filling my cup over and again till the fumes rose in my brain and the room seemed to spin and tilt. With intoxication came the first lessening of the pain that had filled me. I drained the cup greedily, eager for oblivion; by nightfall, when the landlord set lighted lamps on the table and a basket of bread, cheese and olives, I was hopelessly drunk for the first time in my life.

The jug was empty. Marcus called for it to be refilled. We drained it again between us; by then it had become obvious that I at least couldn't take any more. I rose, or tried to, clung to the door-frame, shaking my head and trying to focus my eyes. Marcus supported me, gripping my arm. I shook him off angrily, took three tottering steps into the street. I had intended to relieve myself; instead I fell to my knees and started to vomit. The spasms became more violent; when I had finished I lay full length, cheek against the coolness of the pavement, wishing I could die. Marcus hoisted me unceremoniously, propelled me towards our room. We negotiated the stairs, by some means or another, and I was dumped across the bed. The sense of well-being had departed as rapidly as it had come, leaving behind an ache that seemed about to split my skull. I closed my eyes, drifted half to sleep; when I opened them again Marcus was leaning over me, lighting a wall lamp from a taper. In my confused state it seemed I was back in my old room in Italica. I smiled up at him blearily before memory returned with a rush. All the ills that had beset me seemed condensed into one appalling truth. 'Marcus,' I said, gripping his wrist, 'my mother is dead . . .'

He looked down expressionlessly. 'Yes,' he said. 'I know.'

I was still drunk. For days I had ridden dry-eyed with shock; now the tears began to well. I cried as I think I had never cried before, not even as a tiny child; it was as if a dam had burst, letting through a river of grief.

My recollections of the rest of the night are vague, but one persistent memory remains. It seems Marcus sat by me, wiping my face and throat with some cooling salve as tenderly as ever Calgaca could have done, till the fit had spent itself and I dropped into an exhausted sleep. Though maybe that was an hallucination, brought on by the wine; for in the morning he was his old brusque self, and neither of us ever mentioned the affair again.

We were on the road early next day, cantering north. I still felt giddy from the effects of the drinking bout; but as the sun rose, lighting the rolling hills to either side, my head cleared. I breathed great draughts of the cool air; and it seemed the crushing weight of grief that I had felt lifted fractionally, so that for the first time in days and weeks I could look round me with something approaching interest. I urged my horse alongside Marcus, glancing across to catch his quick smile. His cure, rough and ready as it had been, had worked. The memory of Calgaca was a steady, deep pain; but the formless burden of guilt, that had so nearly destroyed me, was gone.

We had traversed almost the whole length of Hispania. In front of us rose the mountains that separate Tarraconensis from Gaul. We climbed steadily, passing through pleasant upland country set with stands of sweet-smelling pines. Traffic was denser now, much of it moving north; for road after road had joined our route, like the tributaries of a great river. We passed lines of carts, piled high with produce of all kinds, drawn by plodding oxen. In the first part of our journey we had seen no soldiers; now we met a detachment of two hundred or more men. They tramped stolidly, their shuffling ranks taking up half the road. White dust trailed behind them; at their head rode a Tribune, a grizzled, bare-headed man in trews and tunic who acknowledged Marcus' salute with a preoccupied smile. His men were scarcely what I would have imagined for guardians of the Empire. They were bearded and dishevelled;

they kept no particular order, and called and joked to each other in a guttural tongue as they marched. Marcus said they were Franks, and wondered what they were doing this far south of Gaul.

We rested for the night before tackling the final climb. The road wound steeply, clinging to the bare shoulders of the hills. We finally reached the pass through which Hannibal brought his armies in his great march on Rome. At its far end the way dipped steeply once more, and a milepost announced that ahead was the Juncarian Plain. We were in Gaul.

Now Marcus pushed on more swiftly, taking advantage of every hour of daylight. Summer was ending; and ahead, still many days' journey away, loomed the great barrier of the Cottian Hills. We were treading the Via Domitia; we rode steadily, seeing for the first time what war had really meant to the great Province it traversed. We passed plodding lines of refugees, men, women and children; some hauled rough carts piled with their belongings, others sat hopelessly at the roadside, hands extended for charities that never came. Village after village stood gaunt and ruined. Theodosius had left peace behind him, certainly; but it was the peace of a desert.

As we passed through Massilia the traffic, already heavy, intensified. Road after road still joined our route, sweeping in from the vast hinterland of Gaul; each brought its jostling contribution of carts, animals and men. At Segusio, set high among snow-capped peaks, we made a final check of our equipment. The sultry warmth of Baetica was almost forgotten now. I wore three tunics, and the thickest cloak I owned. Marcus was similarly swathed; our breath and the breath of the horses steamed in the thin, bright air.

Surely, I thought, we must soon reach the highest point; but always the road snaked away from us, up and up, losing itself finally in the very clouds. As we climbed, even the normal sounds of the world were stilled. Other travellers tended to bunch together, for mutual protection and aid; so that often enough we camped and moved on alone. Always there was birdsong. The notes fell sweet and limpid, echoing for miles; while the highest passes were filled with a nameless rushing and sighing that was not wind but seemed the movement of the clouds themselves, by some strange magic rendered audible. It was a place apart, laid under enchantment at the start of Time itself.

One morning, just at dawn, Marcus called me from the tent. I crawled under the flap, rubbing my eyes, and caught my breath at what I saw. Around us, above and below, the mountain walls glowed with an incredible hue; a burning pink, like the colour seen at the heart of a rose. As the sun climbed so they blushed and darkened; till the first beams, striking the great rift in which we stood, turned the high rock to silver. We watched without speaking or moving, for maybe half an hour; when we finally turned to the mundane affairs of breaking our fast and striking camp I was filled with a strange sense of peace. It seemed Calgaca's soul, could it have wandered so far, might finally have rested in such a place as this.

The weather worsened through the day, so that when we pitched camp again flurries of snow and sleet drove against the tent. The sleet froze on the road surface, coating it with ice. Next day we led the horses; one false step could have plunged them and us over the lip of rock at our right, into the depths. At midday we came up with a line of heavy waggons, also headed east. There was no question of passing them; they were skidding and sliding, sometimes to within a hand's breadth of the brink. The muleteers levered at the wheels with balks of timber, shouting anxiously to each other; the animals were finally unyoked and led to the rear while the drivers lashed the carts together into one great train. We were over the highest point of our route now; but the descent was if anything more frightening than the climb. Animals and men strained at the ropes, letting the waggons down an inch at a time; we lent our weight to the rest, struggling till nightfall. We pitched camp in the lee of the biggest of the carts, and were glad enough to share the muleteers' fire and supper. Next day we made better progress. The road descended steeply, seeming anxious to be done with the hills. We cantered ahead, reaching at last the gentler plains that fringe the sea.

By this time I was beginning to think of myself as a seasoned traveller. The fleas and lice that had become an inescapable part of my life no longer tortured me; I had acquired some skill in the preparation of food, the porridge on which we mainly lived, and could pick out from the miserable array of taverns those in which we would stand least chance of being knocked over the head for the gold in Marcus' belt. Yet Rome seemed as far away as ever. I was beginning to appreciate emotionally what before I had merely learned from books: the sheer vastness of the Empire. As I rode, so my wonder grew; it seemed impossible that any one man, even my great Hadrian, could ever have controlled it, shaped the course of its history.

We stayed overnight at Genua before joining the Via Aurelia that would take us to our destination. Now there was only one direction. The road flowed south; along it, borne irresistibly, moved what were surely half the people in the world.

For several days we once more skirted the sea. We swung inland finally, through gently rolling hills; and there came a morning when I first set eyes on my goal.

I rode yawning, not long from my bed. Beside me Marcus paced stolidly, impassive as ever, while rain wept from a winter-grey sky. For an hour or more the clustering of buildings round about had been becoming more pronounced. Now they coalesced into a sea of roofs and walls, stately villas set about with gardens, trees and fields; above them, swooping and branching, loomed the aqueducts that fed the greatest city in the world. It was some moments before I realised what I was looking at. Ahead, climbing to meet us, was the pale ribbon of the Aurelian Wall; beyond it the Tiber; beyond again, Rome. She could have stood on seven hills, or seventy; the buildings rose rank on rank, roofs and columns and porticoes, lost themselves in distance under the blowing tides of rain.

It was just as well one of us kept his wits about him, for I certainly lost mine. I charged down the slope of the Janiculum, and would have ridden under the gate had not Marcus called to me curtly to turn aside. Nothing but pedestrian traffic is admitted to the city during the hours of daylight; we lodged our horses at a livery stables instead, and walked back to the Aurelian Gate. Here were real soldiers at last, dressed in the full panoply of the Empire. They stood at ease or lounged chatting to each other, watching the streams of traffic indifferently. We passed through the wall unmolested to the broad suburb that stretches down to the Tiber.

To understand the city in which we found ourselves you must imagine two rings of concentric fortification, the smaller enclosing the ancient heart, the greater an area twelve miles or more round and packed with temples and churches, baths, shops, amphitheatres, tenements, barracks, warehouses, gardens, libraries; everything the mind can conceive. As he plunges into this confused and confusing mass the traveller from any direction finds the road, that has not varied its width by a handspan for a hundred or maybe a thousand miles, finally begins to narrow and wind. It is jostled on either side by taller and taller buildings, as he is jostled by an ever-increasing throng. Here are people from every nation on earth, rich and poor, freedman and slave, black man and yellow and white and every shade between. Barbers and snake-charmers ply their crafts on the pavements; sellers of pies and sausages, sweetmeats, pastry, old shoes, rugs, clothes, yell their wares; money-changers haggle, clinking their piles of coins, lawyers tout for trade. Gangs of roisterers swagger and strut, din rises from building sites, escaped animals career between the feet of passers-by; till the endless babble and noise assault the mind like the sound of a massive sea.

The Rome of my dreams had been stately, white and calm; nothing I had imagined had remotely prepared me for this. For the grimy, streaming buildings, the gutters choked with offal and mud; for the uproar, and the cursing, and the stink. We crossed the Tiber by the Aemilian Bridge. The river raced broad and sullen, its surface pitted by the rain. To either side, dimly visible through the downpour, rose gaunt warehouses; beside them barges, up from Port of Rome, discharged their cargoes on to line after line of wharves. Across the river, beyond the curtain of the second great wall, lay the Forum Boarium, the vast meat market that in those days supplied half Latium. Steam and lowing rose from where cattle, waiting the butchers' axes, milled in their pens. Above, the colossal buildings of Palatine humped and lumbered at the sky; while blood from the slaughterhouses flowed through open channels to the river, an unheeded and unending menstruum. Beyond the Servian Wall the shops began again, their hanging signs lining the streets. Vintners and poulterers, blacksmiths, cutlers, booksellers and corn merchants, vendors of glass and scent, pottery and shoes; I ignored them all, pushing ahead in a species of desperation. There was the Tomb of Augustus, there the Column of Marcus Aurelius; beyond it towered the vaster monument of Trajan. At its foot clustered the Fora; for Rome boasts not one Forum but several, linked in a complicated group on the slopes of the Capitoline. There one can find the Rostra, adorned once with the captured beaks of ships, deserted now by famous men; and beside it the Golden Milestone, set to mark the middle of the world.

I ran to it, breathless from my rush into the city, clapped my hands on its pitted, discoloured surface. At my side, Marcus waited sardonically. 'Well,' he said finally, 'what are you going to do now?'

I stared at him open-mouthed. I hadn't the remotest idea.

Finding the offices of Lucullus Paullus was far from easy. All I positively knew about my uncle was that he was an architect, and that he had premises somewhere in the old City. In my thoughtless way I had presumed he would be famous, or at least well known; but the first dozen people we accosted shook their heads in blank indifference and scurried on. It's an odd characteristic of Romans that they never know a thing about their own great town; they might have lived all their lives in the same sprawling precinct but they can't or won't direct a stranger from one street to the next. Also they are invariably in a hurry. This rushing and scurrying assumes in the end the proportions of a disease. It matters very little to a Roman whether his errand is urgent or not. He will jostle and shove, cursing his fellows and being cursed in his turn; to falter, or break one's stride, is a cardinal sin. We were reduced eventually to jogtrotting beside our victims, shouting our questions into their indifferent ears; some time passed in this way before a fat slave, puffing under a load of cauliflowers, turnips and bread, slowed long enough to favour us with a broad grin.

'I've never heard of Lucullus Paullus,' he said, chuckling. 'And I don't suppose many others have either. But I've a fancy you might mean old Cubicularis. Try just off the Argiletum; between a blacksmith's and a scythemaker's with a big metal sign. I expect it's him you're after.' And he was gone, weaving his way into the crowd.

Marcus and I exchanged puzzled stares. However, we were obviously going to get no more information. We set about finding the place the man had described.

The Argiletum is the street of the booksellers; most of the wealthy publishers have their premises there. The fronts of the buildings on either side are plastered ten feet high with posters and inscriptions advertising the newest works; senators, poets, philosophers, students, folk of all classes, creeds and nationalities, stroll there, stand arguing in the doorways or haggling over the price of manuscripts. The newly rich, anxious to air their culture, get carried there in their chairs; scribes wait patiently by the hour on the offchance of being hired; there too one can usually find prostitutes, Syrian girls in bright, short robes, wrists and ankles flashing with jewels. The street runs from the Imperial Fora; at its far end it dips into Subura, the poor quarter of Rome, loses itself in a maze of tenements, slums, tiny crowded markets. The police avoid it, not without good cause; but I was to come to know it extremely well.

We walked the length of the Argiletum twice without seeing the shops our informant had described. Eventually I spotted them. Between their open fronts a narrow doorway disclosed a flight of rickety wooden stairs. A placard, nailed to the stained wall, proclaimed the premises of L. Sergius Paullus, Architect and Engineer. Below, in smaller characters, was the legend BEDMAKER TO EMPERORS. I glanced at Marcus again, and shrugged. That at least explained the nickname.

We mounted the stairs. As we ascended, a confused uproar manifested itself, became steadily more oppressive. It seemed compounded of many elements; shouting, the thud and patter of feet, thumps and crashes as though bulky objects were being hauled from place to place. I frowned in the near-darkness. There was a second door; I found the latch, shoved. The din enveloped us.

It was like staring into some obscure Hell. The gloom in which the place was wrapped aided the impression. I saw a long, wide chamber, its walls lined with benches littered with wood shavings and tools. Between them, stacked against the walls, piled in confused heaps on the floor, were beds, tables, chairs; furniture of every shape and kind, in every stage of dilapidation and repair. Upwards of a dozen men were scurrying about as if demented, hauling at some stacks, shoving at others, while dust rose in thick, choking clouds.

The editor, it seemed, of this strange pageant stood at one end of the place on a little dais; a stocky, florid-faced man in a dishevelled, once-white tunic, his head adorned by an abominable ginger wig. He was waving his arms as if demented, and shouting in a shrill, nasal voice. 'Citron wood,' he bellowed. 'Citron; nothing but citron, inlaid with ivory. No, gold . . . Abinnaeus . . .' His beady eyes fixed suddenly on the pair of us, still standing transfixed in the doorway. He came bobbing and twitching towards us, weaving his big head from side to side. He reminded me of a boxer in the arena, perpetually ducking and weaving to escape the mailed fist of his opponent.

'Here,' he said, 'you're a useful-looking pair. Get down to the Forum Cuppedinis this instant, with Abinnaeus. I want—no, Abinnaeus has got the list. Abinnaeus . . .' He delved in his tunic. 'Gold piece if y're back inside the hour—no!' Marcus, nothing loth, had reached for the coin; he snatched it back, quick as a snake. 'See my housekeeper after—square meal for both of you. *Abinnaeus . . .!*'

The pace at which he spoke, coupled with an odd, stumbling impediment, made it hard to understand what he said at all, but something in his speech and manner had a bizarre familiarity. The dreadful truth dawned. I said, *'Uncle Lucullus . . .'*

He shrieked, 'Are you both half-wits as well?' Then he leaped, as if I'd stuck a stilus into him; and the tic became positively alarming. 'What?' he said. *'What? What did you say?'*

He interviewed us in a diminutive, equally ill-lit office. Bills, receipts, ledgers, scrolls of paper, overflowed from the shelves that lined it; tables and the seats of chairs were similarly piled. Dust lay thickly everywhere. It looked as if the place hadn't been cleaned or tidied for twenty years, a fact I afterwards found to be true. My uncle read the letter I handed him, twitching and squinting, holding the parchment close to the one dim lamp the place possessed. 'What's Gnaeus thinking of?' he burst out finally. 'Does he imagine I've got nothing better to do than find food and shelter for every waster off the streets who can't turn a penny for himself? And at a time like this. *Fortune and Fever . . .!*'

His manner changed, startlingly. 'This is a critical period for me,' he said. 'A critical period. The next days, the next hours, will decide. I stand at the pinnacle of a distinguished career. But they don't understand. Nobody understands. . . .' He rocked miserably, face in his hands. I had thought earlier the ginger wig looked odd, now I realised why. It was so old that much of the hair had come away in its turn, revealing large patches of the leather base on which it was formed. Indeed he had, as Marcus later observed, a well-shod head.

My uncle controlled himself with a visible effort. 'You will doubtless have heard,' he said in a portentous voice, 'the rumours current in the city of an impending visit? By very important personages?'

So fascinated was I by his variations of character that I omitted to answer him.

'Are you an idiot?' he shrieked, startlingly. 'Where's your tongue, boy? Your *tongue* . . .' His voice changed again. 'There are rumours,' he said unctuously, 'familiar to every person of consequence, that the Emperor Theodosius and his lovely consort are to honour us with a State visit.' His voice sank conspiratorially. 'And I, Lucullus Paullus, having a certain reputation in high quarters—in very high quarters—have received a commission to supply . . .' He raised a finger. '*One dozen beds.*'

It was on the tip of my tongue to ask what even an Augustus could want with a dozen beds at once, but a dig from Marcus silenced me. My uncle saw the gesture; I was afterwards to find out he saw just about everything, despite his habitual air of bemused incompetence. He turned his attention to Marcus, who was grinning broadly. 'Wipe that smirk off your face,' he said coarsely. 'I can tell you right now, there's nothing here for you.' He bounced suddenly from the chair. 'Beds!' he cried, as if some magic resided in the very word. '*Beds* . . . Such beds as the world has never seen. Beds that will epitomise the finest flowering of the civilisation of Rome. Look, look . . .' He grabbed feverishly for a roll of paper. 'Inlaid,' he said, shaking with suppressed emotion. 'Ivory, tortoiseshell, gold . . . while the Royal couches are to bear on their backrests portraits of Theodosius and the Empress, executed in the finest—but look!'

He stooped dramatically, snatched a cloth from a shapeless mound at his feet. Beneath, twin blocks of pure white stone shimmered in the dimness of the little room.

'Alabaster!' breathed my uncle. 'If I told you their cost, if I hinted at it . . . But I am no piddling chairwright. I am Lucullus, Bedmaker to Emperors!'

His face reddened again, abruptly. 'And then,' he said, 'I try to engage a sculptor. A man of reputation, of dignity; one would imagine, of principles . . . An' what does the filthy wretch do?' His voice rose to its former pitch. 'He tries to beggar me,' he yelled. 'Ask a price that would reduce the Augustus himself to scrounging in the market-place!'

I said, 'Wouldn't it have been easier . . .'

'But I am not deterred,' shrieked Lucullus, dancing with passion. 'Screw the sculptor, and all his traitorous tribe. Let the thieving rats squabble with dogs for crusts; I'll do the work myself!'

He took a turn up and down the office, as far as its cluttered state allowed, twitching and mumbling, kicking irritably at the stacks of rubbish that rose on every side. One pile collapsed, sending up a cloud of dust; my uncle burst out swearing afresh, then rushed to me to grip my chin in his fingers. He turned my head to the light, nodding and cursing to himself. 'Yes,' he said, 'you're a Paullus all right, worse luck. . . .' He seated himself again, heavily, squinted at the pair of us and picked up the letter again. 'Form your own opinion of his worth,' he muttered. He fixed me with a calculating eye. 'So you want to be an architect,' he said. 'How much are you prepared to pay me for your training?'

I opened my mouth again, but couldn't for the moment think of anything to say. He rose again, and gripped my arm. His fingers were surprisingly strong. 'Since you obviously haven't got a bean,' he said viciously, 'I'll not charge you for your apprenticeship. But you'll get nothing else from me; I'm not having you under my roof, wasting my food and money. You'll have to make your own way, same as I did.'

As we re-entered the workshop, the din, which had abated considerably, rose once more to its former pitch. At the far end of the place, beneath the one grimy window, an elderly, sallow-faced clerk, who I was later to discover was the missing Abinnaeus, sat working at a desk. Uncle Lucullus deposited me at his side. 'Brother's boy,' he said abruptly. 'Must be mad. See what you can do with him. . . .'

So I began my apprenticeship with my uncle, and four of the most curious years of my life.

My feelings on that first day can probably be better imagined than described. To reach Rome, I had ridden seventeen hundred miles. I had crossed mountains, skirted seas. I was hungry, tired and thirsty, but the suddenness of my induction left me too bemused to argue.

I had been parted abruptly from Marcus; I could only hope that I would see him again. I couldn't believe my uncle meant what he said; that he would neither pay me nor give me board. I was to come to know him better. When I was finally released from my chores—I had been set to smoothing a stack of cheap paper, with ivory combs and a shell—I met him on the stairs. He didn't betray by the flicker of an eyebrow that he had ever seen me before; just stumped past me, head rolling and nodding, and hurried away along the street.

Marcus was waiting for me in the street, still with a broad grin on his face. He hadn't been idle; he was carrying one of our saddlebags across his shoulder and a bulky parcel of bread, cheese and wine. I fell into step with him silently. As we walked he told me he had already found a job for himself, in a big farrier's yard down towards the Tiber, and had secured a room for us. We headed down towards Subura, buildings rising grimly beside us under the lowering sky. He stopped finally outside a tall, dilapidated block of flats. 'Well,' he said, nodding, 'this is home. I got a top-floor room. At least we shan't have anybody dancing on the ceiling.'

I stood and looked round me at the rubbish-choked street, up at the façade of the place. A reek came from it: an old, sour stench of boiled cabbage and dirt. I saw cracked and broken plaster, slime-stained walls. In one place, above a crumbling cornice, grass and shrubs had seeded themselves over the years. The biggest of the stunted bushes, still sporting a handful of decrepit leaves, was indubitably a pear sapling. I shook my head. 'It's a long way,' I muttered, 'from the Forum to the Pear Tree. Not counting the stairs . . .'

That got Marcus. He had always tended to have a curious sense of humour; now he sat on the offending stairs and laughed till the tears shone on his cheeks. 'Sergius,' he said finally, when he could speak again without choking, 'you have the soul of a scholar, I swear it. Rome's going to do you a great deal of good . . .' He rose, and lifted the bundles. 'Come on,' he said. 'Come and see the worst.'

The room we finally reached was tiny, not much more than a pace or two across. It shared with the rest of the house an odour of decay. Rain had seeped through the ceiling in several places, staining the walls, while in one corner the floorboards were so rotten that unless one stepped carefully there was a lively danger of plummeting to the floor beneath. One window, small and square, admitted air and light to the cell. I moved to it, peered through. The view was unexpectedly grand. To the right the cluttered length of the Argiletum ran towards the

Capitoline, crowned with the ancient Temple of Jupiter; across the opposing roofs reared the walls of the Flavian Amphitheatre; to the left I saw clear across the huddle of Subura and the Viminal to the great Baths of Diocletian. Suddenly I realised what the curious events of the day had almost driven from my mind: that I was in Rome, living in Rome, at the heart of the Empire. I turned back, feeling a heart-lifting surge of excitement. 'Marcus,' I said, 'I'm already so much in your debt I shall never be able to repay you. But I promise you one thing. I'll certainly earn my keep somehow, and pay my way.'

'You certainly will,' said Marcus with characteristic abruptness. 'I'm not supporting you, any more than your uncle.' He sat down on one of the beds and began pulling at the straps of the pack. 'That half,' he said, pointing, 'from the end of the bed there to the door, is yours. This bit's mine. Keep the place tidy; you know I can't stand a clutter, and I'm not walking round behind you putting things away.'

'Marcus,' I said, 'what do you think of my uncle?'

He grunted. 'Nowhere near as mad as he looks,' he said. 'But quite mad enough.'

I arrived early next morning to find the office a shambles. The tools were flung about in confusion; one piece of alabaster lay on the floor; the other was cracked clean across, while the slaves were engaged in scrubbing copious bloodstains from the bench and wall. Apparently my uncle's remark about carving the reliefs himself had been meant in grim earnest. Tackling the business with customary vigour, he had soon succeeded in badly gashing his thumb. Undaunted, he wrapped the wounded member and carried on. He rapidly cut himself again; the second injury was so severe he fainted on the spot, and had it not been for the timely return of Abinnaeus, who knew his employer better than I realised, the career of Lucullus Paullus might have ended then and there. He reappeared at the works a few days later, still swearing blue fire. By that time the offending alabaster had been removed from sight; the job was put out quietly, and nothing more was said. In the end the Emperor never came; my uncle took the whole thing as a personal affront, removed the board from his door and swore never to work for royalty again.

I was to come to know him well, too, over the years, but to

this day I don't pretend I ever understood him. Though he styled himself an architect, most of his profits came from projects as widely removed as the repair of aqueduct channels and the restoration of furniture. Nothing was too small for him to tackle, though he took most pride, obscurely, in his manufacturing of beds. The sideline had come about largely by accident. A few years before he had acquired a gang of skilled slaves from an Ostian businessman who had succeeded in bankrupting himself. He had no real use for them at the time, but they had been knocked down for next to nothing and my uncle was a man who hated the thought of any useful commodity going to waste. The speed with which they could produce a bed was nothing short of wonderful; and Lucullus, by virtue of his curious persuasiveness and wide range of contacts, had succeeded in re-equipping half the households of the city. Couches of all shapes and sizes issued from the little manufactory off the Argiletum in a steady stream; and I, who was largely involved in matters arising from minor contracts, spent most of my days immersed in the intricacies of headboards and frames, webbing and cushions. It was an odd fate for a would-be leader of the Empire.

The eccentricities of Uncle Lucullus were many and varied, so varied that he seemed to contain within himself every contradiction imaginable. Normally he was the meanest of men, but he was capable of bursts of startling generosity. He was a shrewd businessman, yet he was often guilty of acts of gross stupidity. One of the stories that circulated about him fits so well with his character that I'm sure it must be true. He had a fine house, on the Viminal; one day he decided he was not enjoying the respect due to his station, and let it be known through Subura that he was willing to receive clients. The following morning his porch was invested by an anxious swarm of beggars, some dolled up in the ancient and degraded toga, all clamouring for the benefits of his wisdom and wealth. My uncle was delighted with the effect, and so far forgot his habitual parsimony as to shower the mob with gold. The following day its numbers, of course, doubled; the night after that the din from the street woke Lucullus in the small hours. The house was besieged by a chanting mob, all hoping for a share of this sudden and mysterious largesse. My uncle, enraged at the loss of his sleep, appeared among them with a horse-whip; he was severely buffeted in the ensuing brawl, narrowly escaping with his life. For weeks afterwards small aggressive groups tended to follow him about in the streets; he eventually shut himself up in his house, refusing to make an appearance till the plebs, traditionally fickle, had found other objects of amusement.

To this misplaced vanity he added an acute hypochondria. He snivelled and coughed his way through the winter months, swathed in a dozen or more assorted tunics and scarves; in the summer he developed violent and I'm sure largely imaginary hay fever. He was also perennially convinced his eyes were failing, and that he would one day go stone blind. That on its own would never have surprised me, for the light he habitually worked by was dim enough to ruin anybody's sight. If, as often happened, the rest of us stayed on after dark he would make a perfect nuisance of himself, stamping round the workshop snuffing out a taper here, extinguishing a lamp there, till he had reduced us to the same Stygian gloom; he viewed each unnecessary flame as a positive financial haemorrhage, and an insult to his business acumen. On grand days he sported a concave emerald, through which he would peer in lordly imitation of the Emperor Nero; other times he would fall into a fit of fretting and wailing, summoning oculists half a dozen at a time to attend him. Then we would spend hours pounding up his salves: the Unconquerable, prepared in accordance with a secret Thessalian formula, the Victorious (as used by the Pharaohs of Egypt) and many more. He would lie back in his chair, damp pads over his eyes, while Abinnaeus, with a set face of resignation, dripped the stuff on to his closed lids. Usually the cure was miraculous; he would spring up revitalised, and curse us all soundly for neglecting our proper work.

Another recurring fantasy was his mausoleum. Death obsessed him; he belonged to one of the richest funeral clubs in Rome and was for ever buying plots of land round the city, on which he swore he would raise a monument rivalling that of Hadrian himself. One of these periodic outbursts took place shortly after I joined him. He called me to his office early one morning, bubbling with enthusiasm; he had, he said, acquired a first-rate plot close to the Porta Flaminia, which after due consideration he had decided was ideally suited for his final resting place. Nothing would suffice but that plans for his Temple of Death be drawn up at once; I would work with him on the building, which would be of weighty but classic simplicity. I hurried to Abinnaeus, highly excited, but the old man merely sniffed. 'If you look in that cupboard,' he said, 'you'll find it's full of the stuff. There must be at least a score of designs, each one bigger than the last. He might settle for one of them, though I doubt it; he usually starts from scratch. He'll sell the land before the month's out, of course; but you'll have to learn, like all the rest. . . .'

I worked hard for a fortnight, regardless. For the first time the endless Orders Abinnaeus had had me learn, the details of columns and capitals, architraves and pediments, seemed to make sense. Whether what I finally envisaged would ever have supported its own colossal weight is highly doubtful; as things turned out, it never had the chance to. A week later, Uncle Lucullus called me sadly to his desk. He had decided, he said—head rolling and bobbing miserably, tears standing in his eyes—that he was unworthy to lie within sight of the great of Rome. Instead he would be cremated and his ashes scattered on the Tiber which would bear them to the sea. It would be a fitting end, full of that humility which had been the virtue of our forefathers. I accepted defeat with as good a grace as possible; the plans, tied sadly with ribbon, were filed away and the whole project forgotten—except that a few months later I found out that just as Abinnaeus had predicted he had received a good offer for the plot from a senatorial speculator with more money than sense, and promptly clinched the deal. Nothing could have been more typical of my uncle.

But such excitements, though not uncommon, barely relieved the tedium of my days. I had not come to Rome to work on sewers, aqueducts and the frames of beds. Certainly I could have changed my lot rapidly enough by joining the army as an ordinary soldier; the Empire was so hard-pressed for men that even Senators had been paying gold to avoid the conscription of their slaves. Marcus stopped me. 'I didn't waste all those years on you,' he said once, 'to see you thrown away in some scuffle with a pack of yelling savages. You're in Rome, you're in a good position to get a commission eventually; until you can make the proper contacts you'll just have to be patient. In the meantime you're learning a trade that's going to be useful to you later on, whether you believe it or not.' One night I pointed out, bitterly, that my life was my own and he couldn't stop me doing what I liked with it. 'Maybe not,' he said thoughtfully, 'but I'd have a damned good try.'

I knew perfectly well when he was serious and when he was not; and he was in deadly earnest. After that there was nothing more to be said; I owed him too much to disregard his wishes and despite the disappointment I knew he was talking sense. However, the business was soon to be driven wholly from my mind, for I met Julia.

Chapter five

Winters in Rome are miserable. Icy winds hoot and whine through the streets; citizens, from freedmen to Senators, are muffled and bad-tempered; everybody one meets seems to have a streaming cold or worse, every building one enters is filled with the throat-catching stink of braziers. Funeral processions wind through the Forum, noisy with trumpets and flutes; the only folk who thrive are monumental masons and undertakers. In contrast, summers in the city are baking hot. The air lies motionless, stifling and thick as a blanket; beneath it the place roars sullen as a hive of wasps. You live with the reek of sewers and garbage-choked gutters; the river falls, baring banks of stinking mud; even the baths bring no relief. They are crammed and noisy, the water warm as soup and tasting of sweat. My first summer in Rome was the hottest Latium had known for years, and tempers rose accordingly. Pointless fights broke out in the streets, running battles with clubs and knives. Once a man was stabbed in our own boarding house. We never found out the rights of the matter; he fled groaning and crying, leaving a thin mulberry-coloured trail, was gone into Subura before the watch could be called. I often wondered whether he lived or died.

Winter and summer, we suffered the nightly thunder of the traffic. Each day, towards sunset, carriages and heavy waggons would begin to line up at the gates. With the last of the light, they burst into Rome like a flood. All produce bound east on the Via Tiburtina, or south into Campania, must pass from Tiberside through the heart of the city; and each driver is obsessed by the urge to get ahead and stay ahead, at any cost.

Nobody will give way; collisions and injuries are commonplace, while it's more than your life is worth to walk along some of the narrower streets. Whips crack, muleteers curse, the din from the iron-shod wheels beats forwards and back between the house fronts. From midnight onwards the wild pace slackens, there might even be an hour or two of quiet; but well before dawn the charge begins again in reverse, empty waggons careering through the streets, racing to reach the western gates before first light. Mixed with them come the raedae and carrucae of the wealthy, back from their all-night feasts, each massive carriage with its train of linkmen and slaves. Chariots, driven with maniac disregard, clatter and bound, horses plunge and neigh. Juvenal once observed that sleep is expensive in Rome; I realised, turning and tossing night after night and waiting for the dawn, just how sardonic had been the poet's wit. Sleep is impossible, except maybe for the dead.

Springtime alone is a delight. The trees in the many gardens burst into their first fresh green; fountains sparkle in the streets; the air strikes fresh and warm. It was on such a fine spring morning, in my second year in Rome, that I crossed the Tiber on an errand for my uncle. Of late, Lucullus had come to rely on me more and more. I was literate, which was more than could be said for any of the slaves; Abinnaeus could read and write, of course, but he had grown too frail and awkward-humoured to do much more than sit and mumble at his desk. As a result I was frequently charged with duties in and about the city, an arrangement that suited me admirably.

On this particular occasion the job was a simple one, a matter of quotation for the supply of lead piping for a villa extension west of the river. I found my destination, the home of the Senator Julius Petronius Gratianus, without difficulty. It was a huge place, occupying a commanding position that overlooked on the one side the whole panorama of Rome, on the other the slopes of Vatican and the distant outlines of the Alban hills. I presented my estimates, spent half an hour haggling with the Senator's Greek secretary before taking my leave and heading back down towards the city.

On the way I walked for some distance alongside the wall that enclosed the grounds. Beside it trees and shrubs had grown up over the years, some to considerable height. I was passing one such clump when I heard a sound I had not heard for a long,

long time: the faint, unmistakable whirr of a javelin in flight, followed by the *thunk* as it struck home. I waited, curiously. In a few moments the sound came again; this time I saw the shaft briefly, glinting in the sunlight, before it plunged from sight at the base of the wall.

I was intrigued. Beside me the nearest of the trees promised easy access. I ran to it impulsively, hauled myself into the branches, climbed till I could see above the wall. I parted the foliage and peered down.

Immediately below me was a stretch of smooth grass, neatly scythed and emerald-green. A few yards from the wall stood a graceful summerhouse; beyond it a screen of bushes hid the little nook from sight of the house. In front of them, thirty or forty paces away, was a girl. I stared at her, and felt my jaw slowly sag.

She was tall, slimly and exquisitely proportioned, and almost wholly naked. Her breasts, high and firm, were concealed by a band of soft leather; another, tied at the sides with thongs, barely covered her loins. Her dark hair, bound by a fillet, had partially escaped; she had thrown the javelin down and was tossing a ball high in the air, running and jumping to catch it as it fell.

Mine was, I suppose, a classic case of love at first sight. In my whole life up to that point I had found little time for women, excepting one; now it seemed on the instant I matured. Dizzying prospects spun in my brain while my senses, rendered preternaturally acute, recorded every detail of the scene. I heard her bare feet whisper on the grass; I saw how her arms and shoulders gleamed with sweat, how above her navel was a little dimple and crease of skin, like the lid above an eye; how the tiny drawers sagged low across her belly, tightened by the sweet thrusting of the Mount of Venus. My body reacted at the sight; I could, to be totally frank, have shafted her on the spot. Perhaps it would have been better for both of us if I had.

She had dropped the ball now and was playing with a little dumb-bell, working all the time towards where the javelin lay in the grass. I watched, incapable of movement, till at the last instant she stooped, snatched the weapon up and threw.

A javelin, coming more or less directly at one's head, is almost invisible. I saw the quick flexing of the shaft in flight, the glint of sunlight running along it, before I lost my hold and fell precipitately from the tree. Above me the spear whispered through the branches and was gone. I sat up to find the vision standing over me, breasts heaving, hands on her bare white hips. I wheezed and coughed, at a loss for words.

'I shall call my father's men,' she said disgustedly, 'and have you executed, or your eyes poked out for prying. Who are you?'

I somehow stammered out my name and business. She laughed spitefully, but what she would have said was never revealed. A voice called from the direction of the house; she turned anxiously, then stooped to seize my wrist. The touch of her fingers shocked, like fire and ice combined. 'Quick,' she said. 'Into the summerhouse, keep down out of sight. If they find you here there really will be trouble.'

Under the circumstances I would have much preferred to stay sitting down, for my condition had survived the fall from the tree. However, I had no choice; I ran after her, stayed crouched in the pavilion till the slave had gone from sight. She turned from me then, seeming indifferent to my presence; she used a strigil on herself, and a towel, before picking up a wrap from the sill. She wriggled into it, hiding the glory. 'Your eyes would certainly have fallen out of your face if you'd stared much more,' she said; then in a softer voice, 'Anyone would think you'd never seen a woman before.'

What pride I had left forbade me from pointing out that I had not. Not in that state at least. I looked down, tongue-tied once more, and saw between my fingers how the mosaic floor of the building repeated the motif of dancing girls, each clad as she had been clad in narrow bands of leather. 'I copied them,' she said. 'I have two more costumes like the one you saw. I like to feel the sun and air on my body when I exercise; do you think that's wrong?'

I could only shake my head. She stared at me, and smiled. She had an oddly sweet smile; to me it seemed to light her face, making it like the face of a child. 'Sit where you are,' she said, 'Sergius Paullus from Hispania, while I fetch us something to eat and drink. After all, you've had rather a nasty shock. I'm sorry about the spear; but you did really ask for it, sitting up there staring and peering like a great stupid crow. Don't go away now.' And she was up, running across the grass, giving me last tantalising glimpses of rounded calves and thighs.

She was back in a few minutes, dressed more normally and

carrying a tray with fruit, bread and milk. We ate informally, she lounging on the mosaic, watching up at me from the tails of her eyes. She told me she was an only child, and frequently lonely, although her father was a great Senator and lived in a palace. She asked me about my own life; I tried to describe my little room in Subura, fell silent when she wrinkled her nose. 'Oh, Sergius,' she said, 'I've never met a boy like you, so shy and sweet. It would be so nice if you could see me again and talk, just as we have today.'

The blood rushed to my head at the very thought. I stammered that I would indeed come again, if only she wished; that I would come every day. She smiled at that, then instantly looked thoughtful. 'I can't always promise to be here,' she said. 'Not at any particular time, I mean. Mummy's so *helpless*, I have to see to practically everything about the house. And you have your work, of course, which makes things worse; I'm afraid it's almost impossible.'

I would have answered that for a mere sight of her I would camp day and night beside the wall, but the words stuck in my throat. Before I could think of anything else she sat up alarmed, opening her eyes and mouth wide. She rose quickly, stood staring through the bushes towards the house. 'I *think*,' she said, 'I *think* my father's home. You must go now; but please don't let anyone see you or I shall get into an awful row.'

I ran to the wall, leaped, and scrambled into the tree. She called after me. 'Do try and come again,' she said. 'I can't promise to be here. But I will try, honestly. Goodbye.' She stooped to pick up the ball, then set off, running swiftly and gracefully over the grass.

I was late back at my uncle's office, but for once his railing fell on deaf ears. In truth I barely heard him. In front of my eyes there floated a luscious ghost. I saw her face again, the darkness of her lashes and hair, the full mouth; I remembered the dress she had worn, the slim gold belt that circled it, how the fabric pulled in little creases round the fullness of her breasts. There was room in my mind for nothing else, nothing at all. I walked through the city that night, hardly knowing where I went. The sun, sinking below the huge bulk of Palatine, touched the sea of roofs with gold, while within the light Julia still danced, shadowy in the brilliant dusk. I watched the nightly rush of waggons into the town, stood above the Tiber to see how the lights of the farther shore sent long disturbed dancing spears into the water. *Julia,* my mind sang, *Julia.* . . . A woman crossed the bridge, walking alone; I thought it was her, but it was not. Her hair, escaping from the fillet, swung and mocked me; I saw her belly again, the delicious blemish of her navel, startling as a blow. The breeze touched my face, smells of the city mingling with the freshness of night air; while the barges moved upstream, driven by sweeps against the slow, full current, each hull starting a vee of ripples that, broadening, flashed back the light before losing themselves in the shadow-confusion of the banks. Rome roared behind me, or sang; I walked on again to where I could once more see her father's house dim against the sky, the roof that sheltered her. I drank wine in a tavern then, well content; that the loveliest girl I had seen had given me drink and food, and asked me to come to her again.

Marcus viewed me with suspicion when I finally reached home. I answered his questions lightheartedly, secure in my secret joy. At night, lying on my bed, the visions returned; I lay silent and dreamy, wrapped in the wonder of what I had found, while the traffic rumbled and crashed up from Subura, shaking the walls with its din. An hour before first light I rose and dressed, ran to see the early market stalls put out their wares. The shadowy doorways, windows from which women beckoned and watched, deserted now, seemed invested with mystery; I sang and laughed, knowing that far across the river Julia lay sleeping or awake, that her breasts rose and fell with the steadiness of breathing, that maybe—my mind was drugged with love—maybe she touched herself beneath the sheets, touched the pricking secret warmth and thought of me. While the first street cries sounded, Subura woke—a part of it, assuredly, had never slept—while the light flushed in the sky and grew and sought the zenith and all Rome waited, poised for the new day. I took home crisp miracles of lettuce and fresh vegetables, new milk and eggs and bread; rough, edible, dark sunlight in the crust of it and flesh. I felt the air against me, liquid coldness rolling on throat and arms, the cobbles under my feet; while strangenesses of light and form awaited me at every turn of the reborn, brightening streets. So God must have felt, on that seventh glowing dawn, looking down on his new world.

A dozen times in the week that followed I nearly confided in

Marcus. Always a deeper insight prevented me. I couldn't have stood his laughter, still less his ribaldry; what I owned was too precious for the light.

It was much more than a week before I saw her again. I crossed the river a dozen times, in stolen hours; a dozen times shinned up the tree by her father's wall, praying to see her there. Once I dropped down into the grounds, ran to the summerhouse hoping to find a letter, a token even, some abandoned toy that might sing of her presence; but there was nothing. Save the dancing girls, shadows of her reality, watching from the pavement with their dark, alluring eyes. I took her a gift, greatly daring; a copy of part of the Aenead, made by my own hand. I left the scroll where she was sure to see. On the second day it lay undisturbed; on the third it had gone. I suffered agonies; but on the fourth day she was there.

She ran to meet me when I peered over the wall. 'Oh, Sergius,' she said, 'how happy you've made me; nobody ever gave me such a lovely book before, I knew when I saw it it must be from you. Do come down, just for a minute; though I can't stay long. Mother has been very ill and fretful; she keeps me by her side the whole time now, which is why I haven't been able to come before.'

To sit beside her, talking to her again, was strange; it was as if before she had had no real existence, as if she was a dream-thing formed in my own mind to which the Gods had given flesh. Her hair, once more, was bound by a fillet; once more it had escaped so that wisps hung darkly glowing, flicked and stroked her satin shoulders as she swung her head. I longed to touch her, feel even the pressure of her hand against mine where she sat beside me gripping the sill of the summerhouse, swinging her legs and laughing. Time passed quickly, too quickly; it seemed only minutes before she was running back as she had run before, leaving me to climb the wall and make my own way home. I watched from behind the bushes till she was out of sight; every movement she made was a precious thing to be loved and remembered, recaptured in the rich anonymity of night.

After that I took to walking the city again, endlessly, sometimes from dusk to dawn. It seemed I was truly alive for the first time; the heightening of my senses persisted, fed by my feverish imagination, so that the most commonplace sights and sounds became invested with new significance. The Pincian and the Gardens of Sallust, where lovers whispered and flitted in the night; the teeming, bawdy richness of the markets; the dead Summoenium where the vast inner wall winds through Rome, casting its shadow on the buildings that huddle at its foot; all these moved me to fresh wonder while the shade of Julia, floating at my side, thrilled and tortured with half-promised delights. Sometimes I would lounge at the city gates, watching the draymen drinking and gambling their days away, the coachmen waiting for trade. Sometimes I walked beside the Circus Maximus, seeing the prostitutes ply for hire, some coyly, some with shrillness and invective; none of them troubled me, and I was glad. The love I felt, that glowed with such fitful heat, would not suffer the defilement of common desire. So the days passed, declining into weeks, and the weeks to months. My life once more had purpose; though what that purpose was I perhaps didn't closely enquire.

I had been working as a scribe in the Argiletum, to pay for my bed and food; I soon found the pittance I earned by my copying was no longer enough for my needs. I took to supplementing my earnings in Subura, in the all-night wine and pastry shops there; in what spare hours I had left I haunted the luxury shops of Saepta and the Forum Cuppedinis, searching for gifts for Julia. I left them regularly now, in the little summerhouse beyond the wall. I gave her bright Egyptian beads, a necklace of carnelian, a pectoral of electrum and gold. Perfumes and face-paint, anklets of twisted greenish glass, a new gold circlet for her hair; once, daringly, a robe of sheer linen that I hoped against hope she would let me see her wear. Sometimes she scolded me for my extravagance, but she always accepted eagerly what I brought. In time, the pace began to tell. I would rise red-eyed and bleary, stagger to my uncle's office after a bare hour of sleep; several times Marcus queried my absences, once took me aside and demanded sternly to know the truth of my affairs. For the first time in my life I lied to him; I told him I was working to buy a commission in the Army, that he would yet have cause to look back on what he had taught me with pride. He grunted disbelievingly, but let the matter drop. A few nights later, as I set out for Julia's home, he tried to follow me; but by that time I knew the city as well or better than he. I lost him in Subura, circled back to feast my eyes on the blank white walls that shielded everything I desired.

But every coin has two faces, and as time passed I found increasingly that my moods of fierce elation mingled with others of darkest despair. Never, my mind whispered, would I ever be worthy of Julia. I, who could barely afford a rabbit-hutch in Subura; how could I ever build her a summer palace on Capri? For that, and that only, seemed worthy of my love. Despite myself my imagination still ran free. I saw the lamps that would light the place, glowing against the summer blueness of sky and sea; the fountains with which I would fill it, the pools and scented trees. In the pools bright fish swam, great flowers opened their petals to the sun. Sometimes Julia bathed among them, languorously, attended by slaves scarcely less lovely than she; I saw her white breasts and arms, her hair as it trailed her shoulders, floated free like fronds of luscious fern. Sometimes I told her my fantasies, thrilling at the dream reflected in her eyes; sometimes the visions came to me at night, haunting and bewildering, mingled oppressively with the roar from the streets. Julia waded from the water, belly gleaming, while I turned and tossed in burning need, groaning aloud with despair, hard as the Pillars of Hercules and with no hope of relief, unless I performed that sacrifice to Venus that left me shaky and dull-eyed, turned love to a wilderness of self-contempt.

Sometimes Marcus swore at me in the dark. Once he threw his boots.

Julia herself was a capricious creature, full of opposites and contradictions. She explained them as inherent to her sex, the mystic ties women have with earth and moon. Sometimes she would appear as innocent and sweet as on that first day; at others she would taunt me with my lack of connections, asking me how I would ever set about becoming a famous man. Once she took my hand, demanding to know how I came by my disfiguring wound. I told her truthfully that I had got it in a duel. She mocked me, affecting not to believe, saying as a baby I had been trodden on by a horse; she pretended to see in the curving scar the very mark of its shoe. Next day, though, she seemed lost in admiration, demanding I tell her the tale again and again, marvelling at my prowess and bravery. 'You're quite obviously destined for great things, Sergius,' she said.

'You'll evidently become a soldier, perhaps surpass even your patron.' For little by little she had drawn out all the details of my life, including my boyhood love for Hadrian. She sighed, leaning back on the grass. 'That's the sort of man I always wanted to marry,' she said. 'A soldier, riding into Rome in triumph; who would let me share his chariot, wear golden armour and a sword. Be a soldier, Sergius, if only to please me.' I basked under her praise; but when next we met she brought two simpering servant girls, who to my fury would not be driven away. Our intimacy had emboldened me; but when I asked her to dismiss them she turned on me furiously. 'What a thing to say,' she said scornfully. 'Sergius, you must be mad. What are you hoping to do with me, when you get me on my own? To see how I've trusted you, as well; what a thing to say!' The girls giggled behind their hands; I coloured furiously, tongue-tied with anger and shame.

At such times the despair would grip me more strongly than before. I would stamp beside the Tiber, face set, staring down at the endless yellow racing of the water. One day, perhaps soon now, a wealthy suitor would present himself at Julia's gate; and there would be an end of hopes and dreams, of life itself. Sometimes it seemed I should fling myself into the river, have done then and there with loneliness and pain. It would be better for Julia, I told myself, to finish the thing for ever; I was certainly not fit to step within her shadow, my continued presence could only bring her grief. Perhaps too—I hated myself for the thought, but it would recur—perhaps she would keep a corner of her heart for me; stand and watch the river that had borne so many sons of Rome down rolling to the sea, and shed a tear for what had been and what could never be.

Spoken thus baldly my passion seems absurd, even to me; but when the mood is on me I can still relive the burning hopelessness of those years. Sometimes, Julia seemed to understand. Then she would suggest, with touching shyness, that all might not be really as black as I feared. Love, she would say, was the most important thing in the world; love, that conquers all barriers, sweeps aside all opposition to its schemes. If only love is strong enough, what may a man not achieve? Why, already I was becoming known in Rome; did I not speak with Senators on equal terms? I must work, tirelessly and unceasingly, and the future would resolve itself, that I would see. Occasionally the promise would be even more firm. Once, as I climbed my inevitable tree, she ran to me happily, calling and waving. 'Look,' she said, 'today I'm a Spartan girl, the equal of any man. Don't you think it's lovely?' And she turned to show me how her robe was open at the side, baring her hip. I moved towards her, impulsively, but she darted out of reach. 'No,' she said, 'sit down and behave yourself, Sergius Paullus, or I shall go away. Go on, I mean it; sit down, exactly where you are, and promise not to move.' I did as I was told, dumbly; and she crept forward, an inch at a time, till her mouth was a bare fraction from mine. My heart thundered; I saw her lips part; then her tongue was pushing, her teeth were jammed against mine.

'I always kiss like that,' she said demurely. She was sitting on the grass, weaving daisy chains; splitting the stalks with her long, filed nails, slipping them each through each. 'Or I don't kiss at all. You're so sweet, Sergius, and have been such a friend to me. I hope you didn't mind...'

She wanted a baby leopard, or a cheetah to walk with her on a lead; but such as came to Rome were consigned to the arena, their prices above anything I could afford. I bought her a monkey instead, a tiny plaintive-faced thing, scarcely bigger than my hand. I left the cage on the sill of the summerhouse where she was sure to see. For two days after that I was kept too busy to cross the river. When I went again the cage still stood where I had left it, the little creature dead in the bottom of it. Round it on the grass the flower chains she had woven lay shrivelled and brown. I took the thing away and buried it, and said no word at all.

Looking back it seems strange how the great events that were shaping themselves in the world could leave me so unmoved, but perhaps that is always the way with men. I have heard folk claim the moon is a mighty plain, as broad as the entire earth; yet a leaf of a tree will hide it. So it is with us; the near obscures the far.

I lived in a frightened city. Rome was as she had always been, to the outward eye; bustling, noisy, vibrant with life. Grain-ships docked at her wharves, unloading their cargoes in golden streams; her aqueducts still delivered their countless gallons each day; men rose in hope, bedded in despair; and the arena roared. Emperor after Emperor, from the great Constantine onwards, had tried to forbid the games, but the people still would have their blood and bread. The Church railed, seeing in the amphitheatre the very playground of demons; but to no avail. Scaevola still thrust his live hand into the coals, Icarus took his fearsome flight, Orpheus bleated while his guts were ripped by bears. Under it all lay terror; the terror of a hectic, swirling life that soon must end. I see it now so clearly; how in the quiet watches of the night the eyes of the city turned north and east, where in a great ring stretched the Provinces that for years had guaranteed the safety of the hub of Empire. Not so many years before the plague from China, the Huns, had swept most of the Gothic tribes into confusion, so terrifying the survivors that the Thervingians, themselves no mean fighters, had begged to be granted asylum within the Empire. Valens, hard-pressed for manpower, had agreed, to the horror of most conservative thinkers; and an entire nation had been ferried to the imagined safety of Thrace. I remember my father remarking more than once that that single act could yet spell the doom of the West. For twenty years now the tribesmen had rested on Rome's flank, uneasy and turbulent neighbours. They had risen once already; and at Adrianopolis Valens had paid for his short-sightedness with his life. The rot had been checked savagely. A pay parade was called, throughout the Army; after it, no Thervingian recruit remained alive. But the act, necessary though some thought it at the time, had left a smouldering residue of hate. Thrace was notoriously infertile; more than once Alaric, hereditary King of the Goths, had applied to move his people to Illyricum, and been refused. None knew when the banked fires might burst into fresh fury; and Romans at least had no illusions about the result. Under Theodosius the balance of power had been shifted radically eastward; Latium, rich and comparatively unguarded, could be invaded by a yelling horde if Alaric chose to swing his armies round the Adriatic.

Yet there were new forces in the world, new names on which to fasten old hopes. Foremost of them was Stilicho, barbarian and Magister Militum. He had, so the rumour went, accepted the hand of the Emperor's niece, tieing his fortunes to the fortunes of the West; and he was without doubt the finest soldier of the age. In Marcus' view the fact that he was a Vandal showed the depths to which the Empire had sunk. I disagreed with him. Rome had always leaned heavily on satrapies and allies; if Stilicho, or any other barbarian, could ward off the dangers that beset her then he had at least my undivided support. In any case, as I have said, I troubled myself very little over international events. To me, the flicker of Julia's lashes could send more kingdoms crashing than the Huns.

In my second year in Rome news reached the city that another of my father's prophecies had been fulfilled. In Gaul, Arbogast had rebelled against his young master, murdering the boy and setting up one Eugenius, a misplaced Professor of Rhetoric for whom the illiterate Magister Militum seemed to have profound respect, as Emperor of the West. The development caused more gratification than anything else in official circles, for most of the great Senatorial families had remained obstinately pagan, and there seemed a real possibility that if the new Augustus succeeded in imposing his will on the West the ancient Gods of the Empire would once more be restored. Rumour had it that Symmachus, wealthiest and most vociferous of the opponents of the Church, was already planning games in honour of Jupiter Optimus Maximus; sacrifices were once more offered on the Capitol; Christians, and by this time there were many in high places, looked anxiously to the Emperor for reassurance. It wasn't long in coming. In a lashing rescript, Theodosius forbade all forms of paganism. Offerings to images, even the lighting of lamps, could lead to fines and banishment; the hanging of garlands would be punishable by confiscation of property, while divination from entrails was declared treason and became a capital offence. Thus rebuked, the city subsided sullenly; the warning finger could not, after all, have been more plain. Damasus, Bishop of Rome, declared a day of thanksgiving, while a hastily assembled Senate passed a resolution of loyalty to the Emperor and his ideals. Not, as Uncle Lucullus remarked with a flash of my father's sarcasm, that it greatly mattered any more what was tattled round the village pump; loyal or disloyal, neither Theodosius nor his family seemed to have the slightest interest in restoring the city to its rightful position as head of the State.

Theodosius was slow to move into Gaul. Rumour had it that he was a sick man, perhaps a dying one; certainly two seasons passed before the field army mobilised for the long trek to the West. For a fortnight Rome echoed to the marching tramp of levies; patriotic slogans appeared on the walls while orators declaimed from every other street corner, some urging the citizenry to remember past glories and honour the State with their arms, others insisting that now was the time to strike off the fetters with which Constantine had bound the Empire. Many Pannonians of doubtful origin had, of course, acquired estate under Valentinian; now they grasped at the opportunity to rid themselves of the fear in which they had walked ever since. Even the old spectre of Republicanism was raised. Rome's walls were strong, they whispered, her gates impregnable. Let her but resist the Christian tyrant and she would see the West rally to her side; she would be again what she had been before, the mistress of the world. The propaganda was not without effect. Angry mobs of partisans roamed the streets; fighting broke out in Subura and some twenty shops on the Aventine were pillaged and burned. The reaction of the City Praefect was sharp. Troops and police were drafted into Rome and proclamations appeared on all principal buildings promising the death penalty to disturbers of the peace; there was talk of a curfew being imposed, and threats were made to restrict the public supply of grain and wine. What the end of the affair might have been is difficult to say; in the event, the trouble was terminated by the arrival of an Imperial messenger escorted by a troop of hand-picked comitatenses. The Emperor was victorious; Arbogast and his puppet had been destroyed, and Gaul once more secured.

Relief in the city was nicely compounded with bewilderment. Nobody knew quite what would happen next. Theodosius' reign had for the most part been characterised by tolerance and humanity, but lately his attitude to his opponents had been hardening and he had shown in the past that he was capable of taking appalling revenge. The Senate, thanks partly to Valentinian's careful spadework, had always solidly disapproved his policies; now the old fear of a purge was strongly revived. Also the victory was said to have been largely due to a sudden storm of wind, which had scattered and confused the forces under Eusebius; Bishop Damasus lost no time in proclaiming the miraculous nature of the intervention, and many waverers, convinced by such a striking demonstration of Divine wrath, moved hastily into the Christian camp. The opponents of the Faith rejected the notion with renewed bitterness, and uproar seemed about to break out once again. Meanwhile the City

Fathers, convened once more by the overworked Praefectus, passed a motion according to Theodosius the gratitude of the Empire and requesting that the Augustus honour the capital with his presence at the Triumph voted to him as saviour of the State. The answer came swiftly. The field army was on the move; the Emperor was already marching on Rome.

The city that a few weeks before had been industriously preparing to celebrate the final victory of paganism threw itself with fresh fervour into the new project. Under the circumstances it would not do for anybody to be found wanting in protests of affection. Senators, including Symmachus and his noble cronies, vied with each other in preparing costly entertainments; streets were swept and scoured, the Royal apartments once more hastily prepared. Folk began to flock in for the spectacle; after them poured quacks and mountebanks, prostitutes, cut-throats and thieves, all the riff-raff of Latium. Every lodging house was crammed to capacity, food and wineshops did a non-stop trade. The Praefect let it be known that the event would be celebrated by games, and by public banqueting on a hitherto unheard-of scale. When the van of the approaching army was reported within a day's march of the city, expectation rose to fever-pitch; but the intelligence, exciting as it was, scarcely left an impression on me at the time. My own private drama had at long last reached its climax.

When news of the Emperor's victory reached Rome, my uncle, who till that time had shown no interest at all in the doings of either Theodosius or his House, chose the occasion for a spectacular display of grief. Things were far from right with the world, he proclaimed; nor would they improve till the Altar of Victory, so impiously removed from the Senate House by these same upstart Christians who now paraded their triumph in the streets, be restored to its rightful dignity. His despondency, typically, took an original and ingenious form. He retired to his bed, where he informed us he intended to stay till Rome returned to her senses and the proper worship of her ancestors. In the meantime he kept the chamber scrupulously darkened, and gorged himself on pease-pudding. Till the fit passed there was nothing to be done; I took over the daily affairs of the office, and it was a fortnight before I could cross the river again to Julia's home.

I had decided, perhaps unreasonably, that I could stand the

situation no longer. For two years she had seldom been out of my waking or sleeping mind. Meanwhile my situation had considerably improved; parsimonious though my uncle might be, it seemed certain I would one day inherit some at least of his business interests. I could in fact already have set up in some capacity of my own, had not an innate honesty prevented me; now Julia must answer the question that had come to mean so much to me. The gulf between us was wide, perhaps unbridgeable, but no one could have given greater earnest of his love. I must know, finally, if I had any cause to hope.

I accordingly sent a message, by a slave, that I must see her without delay. I chose the very day Theodosius was due to make his triumphal entry, for I had learned that though Petronius, as a prominent citizen, dare not absent himself from the Curia, he had expressly forbidden his family and household to attend the celebrations. Rome was already in tumult as I crossed the Tiber. Churches and public buildings were gaudily decked with banners, many bearing patriotic slogans; drink and sweetmeat sellers circulated in droves; every inch of the processional route was lined by a noisy and jubilant throng. Police and soldiers bawled themselves hoarse trying to keep some sort of order. Grandstands had been improvised in many places, while every window and parapet overlooking the Sacred Way was crammed with spectators; both householders and public officials had been charging exorbitant sums to admit sightseers to their premises. The sun beat down strongly; I shoved and barged my way through the mob, and was already sweating profusely before I reached the river.

The silence of the Janiculum was almost startling by contrast. Most of the houses were closed up for the day, their owners and staff down in the city. I dropped my pace to a walk, dawdling uncertainly. Once I nearly turned back, but Julia would be waiting. I had gone too far to withdraw.

My heart was thundering against my ribs by the time I reached the villa wall. I leaned my palms against the sun-warmed stone, trying to calm myself. The hubbub from the city reached me faintly, drifting across the roof-tops; above me a yellow lizard flicked his tail and was gone into a crevice in the twinkling of an eye. I sighed, and walked heavily to the tree. That a prospective suitor must still gain entry by such means struck me for the first time as bizarre.

I peered over the wall, half hoping the summerhouse would be empty. It was not. Julia sat with her arms folded and lips compressed, tapping impatiently with one slim foot. As I watched she rose, took a turn across the little arbour, plumped herself back down bad-temperedly on the sill. I called softly, dropped down inside the wall. She looked round scowling as I approached.

I paused at the threshhold, involuntarily. Never had she seemed more desirable. Her hair, freshly brushed, was garlanded with flowers; she wore it loose, in a dark cascade that fell nearly to her waist. She was wearing the gauzy dress I had bought for her so long ago. A girdle, tied at her waist, accentuated her slenderness; her breasts showed clearly, their red buds pricking the thin cloth. She was breathing heavily and her cheeks and forehead were flushed with rage.

'About time too,' she snapped before I had had a chance to speak. 'Oh, my stupid, stupid father. . . . Just listen!'

A distant wave of cheering swept up from the road.

'Everybody will be there,' said Julia. 'Just *everybody*. . . . But I've got to sit here, because of that ridiculous Symmachus and the rest and what they said to father. Their families will be going, I *know* they are. . . .'

The time seemed hardly propitious for what I had to say, but once having nerved myself I had to go through with it or never speak at all. 'Julia,' I said, 'please don't be upset. I must speak to you; I have something very important to ask.'

'Have you indeed?' she said. 'Well, I've got something to say to you. So just sit down, and listen to me first.'

I sat unwillingly, full of my own plans and wondering how best to broach them. She gathered the gown modestly round her feet, sighed; then told me, quite blandly, the news that broke my world apart. Her father, she said, was tiring of city life, and with the growing uncertainty of the times had decided to remove to Neapolis, near which he owned an estate. The household, including Julia, were to travel within a few days; the Senator would follow when he had concluded his affairs in Rome. So, sad though it undoubtedly was, she had come to say goodbye; we would certainly never see each other again.

At first the magnitude of the tragedy left me at a loss for words. I could only sit staring at her, repeating stupidly, 'But you can't, you can't. . . .'

She came to me then, knelt to take my hands. 'Sergius,' she said, 'please try and understand. I'm sorry I was rude just now, truly. It's been lovely all this time, knowing I could come here and meet you and that you would be a friend. I love you truly, as a friend, and will certainly never forget you. But can't you see it's over now? I couldn't possibly influence my father, once he's made his mind up about anything it's hopeless. And in any case I very much want to go south, I'm sick and tired of Rome. If you get angry you'll make me most unhappy.'

For me that was the last straw. For years now I had thought of nothing but her slightest whim; everything she could possibly need I had toiled to buy for her, but in return she had given nothing. I saw in a flash what I had known all along but refused to admit: that I had never mattered to her, that she had been playing nothing more than an amusing game. I pointed to the wall, nearly choking with rage, told her how easily she could climb it and be free. As I had climbed it, time after time, till the very branches of the tree beyond had been worn smooth by my feet. Now I was to leave her, slip quietly away, taking particular care to cause her no pain. My family, I told her, had never been noted for philosophy; to follow such a course needed more will than I owned. 'Julia,' I said, 'I love you more than any man ever loved a woman before. Please, please don't do this to me; don't try to make me go away. Please say one day you'll be my wife.'

Her reaction was startling. She snatched her hands away, jumped to her feet and began to stamp with temper. This, she said, was what came of all her efforts at friendship. From the first she had been sorry for me, sorry and that was all. She had offered me companionship and love; now I couldn't even respect her breeding. It was impossible that a member of her family, the richest clan in Rome, could ever mix blood with a commoner; for she had only my word that I was even an Equestrian, and certainly my current behaviour was doing nothing to support the claim. She had even put on her prettiest dress to say goodbye to me, the dress I had bought and that she treasured, so that I could see her in it and take away a lovely memory; and look what the result had been. I had insulted her; now I could leave, at once. Or she would call the housepeople, and have me slung into the road.

Appalled though I was, her outburst precipitated a worse rage than ever. 'You wore that dress,' I shouted at her, 'for one reason and one reason only. To drive me wild. As you've been driving me wild for two whole years, leading me on with promises of this and that that had nothing on earth to do with friendship; and that you know perfectly well, as well as I. You're a liar, and a cheat, and if you had more courage you'd be a whore; and to see me floating down the Tiber, that you'd class as a victory.'

The horror that filled me, as soon as the words were out and unretractable, I remember to this day. All was quite obviously over between us, but before she left the spot I would give her something more permanent than a memory. Some of that Patrician blood of which she was so inordinately proud would be spilled forthwith; I had been played with more than long enough.

Obviously she saw the intention form in my eyes. She turned to bolt, but for once she wasn't quite quick enough. I grabbed for her, caught the dress instead. She tripped, rolling across the grass; and the whole thing pulled from the clasp, came away in my hands.

Beneath it she wore nothing but a little loincloth, scarcely larger than the garment in which I had first seen her and drawn tightly between her thighs. The sight inflamed me. I flung myself on her, one hand between her legs squeezing and kneading at what I found there; and my stilus, the bronze spike I always carried in my belt, flew out and landed on the grass. She grabbed for it, fast as a snake, lashed up at my face. I flung myself away; and she changed her aim, stabbing savagely at my hand where it rested on the ground. I felt a blow, a dumb and painless shock, and stared down amazed. The little weapon, needle-sharp and driven with all her force, had passed completely through my hand; the tip protruded a clear inch from the palm. Twin streams of blood were already running from the wound, uniting at my wrist to drip to the grass by her feet.

I sat back, stupidly. It seemed that horror and surprise had driven all sensation from me. I remember thinking absurdly that this just would not do; I could hardly go about with a metal spike poking through my palm. I gripped the handle of the thing, tugged till it came free with a creaking against bone. I had thought the wound would close; instead blood spurted brightly. Fear came then in a great flood, and blackness at the edges of my sight; and I was beating her where she still lay on the grass, and she was screaming at the full pitch of her lungs.

I think I might have killed her, but a shadow flicked across me and some instinct, gained maybe from those years of training with Marcus, made me roll violently to one side. The slave overbalanced, grunting; the hand-sickle he carried, that would certainly have buried itself in my back, wasted its stroke on the air. He was up instantly, coming at me again. Injured as I was and unarmed, I was in no condition to face him; but there was no time for decision or conscious thought. My legs were already pumping, carrying me to the wall. He was close behind as I gained the top; the hook swung, missing me by inches. I fell heavily from the tree, rolled to my feet and set off for the city at a stumbling run, trying as I went to staunch the wound with the torn edge of my tunic.

In my bemused state I was convinced I was still pursued, that any moment the blade would come slicing at me again. The sun burned on my shoulders and back, dazzled from pavements and reeling white walls. I wept and cursed, calling on Hadrian and Marcus, Julia, Calgaca; while ahead the roaring grew, mingled finally with the pounding in my ears. Fear and grief robbed me of what sense I owned; my one thought was to escape, to my bed, to my room, lie and lie in silence and let the sickness go away.

My chest was hurting. I was aware, dimly, that my legs still pounded, but I could no longer feel the pavement beneath my feet. Round me the din seemed solidly to fill the air; there were eyes and hands and faces, shouting mouths. The pressure of bodies was immense; it was as if I ran not in air but some thick, impeding fluid. Time after time I thought I saw my own street in front of me, but all streets were the same. The blackness was back, leaping and flickering. I reeled, like somebody drunk; and the pavement rose, quite slowly it seemed, to punish my knees. The hands found me instantly, and the faces. I wrenched away, and thought I was running again; but this couldn't be, for the paving still pressed my shoulder and side. I squeezed my palm between my knees, trying to stop the huge, hollow ache.

Cloth flapped before my eyes, poppy-bright. They had my wrist, they were trying to lift and turn. I was raised, unresisting. Above me I saw the roof-edge of a building, dark against the hot blue sky; and a statue on its column, white face remote, apple-small and blazing like a god.

The noise, though still immense, seemed more remote, like sounds heard in a tunnel or great vault. Also I was shivering; I was glad when somebody draped a cloth round my shoulders. I sat indifferently, seeing between the legs of the crowd the swirl of wheels and hooves, feeling the shaking of the ground where Theodosius Augustus passed in glory.

Chapter six

The Emperor dealt leniently with Rome. For days the field army camped under the walls while the city accorded him the honours that were his due; at the end of that anxious time of waiting he made his intentions known. Almighty God, he told a packed and silent Senate, had once more manifested his divine will. By his hand the Empire's enemies had been scattered, the doubters silenced for ever. Therefore let all men praise the Lord for their deliverance. Meanwhile, as a reward for faithfulness, he named his younger son Honorius as Caesar of the West. For Rome herself he had prepared a unique honour. Other cities might surpass her in the arts of government, but from henceforward her Bishop would be foremost in the councils of the Church; through the good Damasus her voice would once more be heard in world affairs. How Symmachus and his party received the news was not recorded; for the moment it was enough that Theodosius departed, grey and worn, for Constantinopolis. There less than six months later he was to die, leaving a child of thirteen the master of our fates.

Shortly after the Augustus and his followers left Rome the guards at the Aurelian Gate witnessed a curious sight. A singular traveller passed through, headed away up the long slope of the Janiculum. It ill became the dignity of such a man to walk; accordingly he had hired a bobtailed mule, across which he sprawled ungracefully, muffled in a variety of tunics and scarves although the day was warm. He wore an ill-fitting ginger wig; over it, to protect himself from the dangerous glaring of the sun, he had clamped a floppy-brimmed straw hat. From time to time he sniffed, rubbing his nose with the back of his hand or blowing it vigorously into a large and insanitary kerchief. Some natural infirmity, coupled with the broken-down gait of his mount, imparted to his movements a peculiar bobbing motion; it was as if he ducked and weaved incessantly to protect himself from the blows of an invisible assailant. The guards, who were used to oddities, watched the voyager stolidly out of sight before returning to their contemplation of the sailing clouds and sky.

Petronius Gratianus received my uncle courteously. Wine was produced, a sweet mulsum, and trays of nuts and confections. A lesser man than Lucullus might have been subdued by the opulence of his surroundings; my uncle pecked and gobbled, seeming oblivious to the material evidence of wealth, while the conversation wheeled ponderously round to the purpose of his visit. Some of that conversation I heard afterwards from Marcus; the blanks I can fill in readily enough for myself.

The Senator was affable to a degree. Certainly the affair was regrettable, unfortunate in the extreme; but as it happened no lasting mischief had been done. He himself was prepared to overlook the whole business; after all, he had enough to engage his energies without troubling his head over the fate of the odd trespasser in his grounds. And—here was the point— the boy had offered violence to his daughter; he could still find himself dragged most unpleasantly through the courts, were Petronius so inclined. Thank the Gods the girl had had enough spirit to defend herself; the intruder—pardoning Lucullus' presence—had, after all, only got what he deserved. No, there had been upset enough. Daughter hysterical, wife in tears . . . he was prepared to drop the whole matter, forget it. Indeed he insisted on it. The Senator rose, giving Lucullus a decent cue to leave.

My uncle, who had finished an entire bowl of sweets, looked round mournfully for more. None was forthcoming. He stared at the Senator regretfully. His twitch had become worse, as it always did in moments of high stress. His head, large and irregular-featured, rolled alarmingly; his speech thickened and blurred; but Petronius was uncomfortably aware that throughout the performance two eyes like beads of steel remained clamped unwaveringly on his face.

Court cases, agreed Lucullus, were unpleasant in the extreme. But while the subject was in the air . . . the boy was sick, might easily die. In that case some action . . . almost inevitable. Endless repercussions. Manslaughter, one might almost term it. Perhaps even . . . murder . . . The word seemed to hang between them in the air, unpleasantly.

The Senator relaxed, gracefully. Obviously the situation stood in need of further clarification. What exactly did the good Lucullus have in mind?

Very little, really. This was a social visit, purely social; the Senator, of course, appreciated that. A mere chat between friends. Regardless of its outcome, the business could easily have an adverse effect on the reputation of the household. Rome, unfortunately, was a city of rumours. Should a certain tale, for instance, be carried through the streets of . . . say, for the sake of example, Subura. . . . The girl . . . beautiful girl, the Senator's daughter . . . waiting to receive . . . ah . . . *lovers,* beside the wall . . . Results difficult to assess. Could bring troupes, positive regiments, of undesirables . . . house besieged . . . the retreat to Neapolis could yet become something in the order of a rout.

All in all, a quiet settlement was to be preferred. A little favour, between friends. The lad was useful certainly, but headstrong. In need of discipline. After all, he'd never make an architect. Heart not in it. A new start perhaps, at some little distance from Rome . . . My uncle shrugged exaggeratedly. The farther away, he seemed to be suggesting, the better.

The Senator understood, perfectly. And agreed. After all, for a man with his connections the affair would be easy to arrange. Light broke across the curious battlefield; and Lucullus rose, with all the dignity of the Fathers of the People themselves, hobbled back to his round of eyedrops and cough mixtures, pease-pudding and beds and chairs.

Sometime, I opened my eyes. Marcus was with me. His face seemed to swim in sharp light, though the rest of him was shadowy and vague. I remember thinking inconsequentially how old he had begun to look; the lines deepened on his face, hair at his temples iron-grey and white. 'Well,' he said, 'you've got what you've been asking for. And I hope you're satisfied. If that becomes infected you'll get a damn sight more.' He turned then, I saw, and left the room, tramping heavily away down the stairs. It was the last thing I was to see clearly for some days.

By midnight the pain in my arm had worsened; by dawn—
or maybe the third day or the second, I lost all real count of
time—it was intense. It seemed I held the sun gripped in my
palm; when I raised my arm, with a creaking of great doors and
bones, his hot juice ran to the shoulder, leaving behind streaks
and traceries of bright-coloured flesh. The pain heightened by
degrees. It was as if I climbed a range of hills; each plateau
gained, which was a resting, however slight, from agony, led to
a worse steep; each steep took me farther from comfort and
help, from what I had known as life. My senses told me
Marcus was close; but I could neither speak to him nor understand his voice, he seemed separated by great gulfs from where
I hung or floated. I have wondered since why I came so close to
death. I was young, and healthy; there seemed no cause for the
poisoning of the wound. Except that as I have grown older I
have come to believe the mind and spirit, the insubstantial
parts of man, to be intimately connected with the flesh. If the
soul is sick, no potion or medicine will be of use; the body will
also pine.

Certainly in my lucid moments I saw no reason for continuing to live. I searched my memory time and again recalling
every meeting with Julia, every word that had passed between
us. Could she have lied so persistently? It seemed absurd;
what on the wide earth had she stood to gain? Where then
had I sinned, where lay the fault that had led to such bitterness? I could find nothing; in all the time I had known her I
had remained faithful to my dream. So it seemed my crime lay
simply in being poor; an eccentricity no woman, least of all a
Roman, was ever likely to forgive. It was I who had tried to
take everything, giving nothing in return. When I remembered
what I had proposed I cringed with self-contempt. I would have
dragged Julia from home and family, made her share my
poverty in Subura; for what? I saw myself clearly for the first
time; bumbling and clumsy-bottomed, shod like an Hispanian
colt, whining and scrabbling for things outside my reach. And I
had dared to feel rage at her refusal! Sweat stood on my forehead at the very thought. Then the agony would come again
and I would wonder why I had not seen before that Julia, my
griefs and sad estate, were shadows. All that mattered, all that
a man should ask, was that there be an end to pain.

Marcus fetched a harassed Precinct doctor, practically lugging him up the stairs by the scruff of his robe. What he did to call me back to life I can't say, but I remember lying sleepily, without pain. Rome, it seemed, was calm in evening light; a mellow light, very golden and clear. The noise of the Circus rose over the city, like the cheering of a distant sea; above, many Gods floated in the crystal sphere of the sky.

I woke, in our room; a square, cold little room, smelling of cabbages and rain. The raindrops banged and clattered, bouncing from the one high sill; others tinkled from the ceiling, falling into buckets and an earthenware bowl. As my senses adjusted, the noise receded. I would have drifted back to sleep, but the persistent shaking of my shoulder roused me. Marcus was leaning over me. He seemed pleased by my recovery; though why he should roar with laughter and rattle papers underneath my nose was more than I could understand. He let me rest, finally; later he read the words to me, again and again, till my brain could take them in. I reached for the things then, dimly. Everything was in order; here was my posting, countersigned by the office of the Praefect of the Gauls, here the chit authorising me to draw my kit and stores, here the pass entitling me to levy horses at staging posts throughout the West. I examined the seals again, too confused for thought. My life had been changed once more; I was to serve Honorius Caesar as a Tribune in the Army of Rome.

Two days passed before I was strong enough to demand of Marcus how the miracle had come about. He was gruff and evasive; he told me, bluntly, not to look a gift horse in the mouth. I persisted until he admitted, I suppose in self-defence, what I wanted to know. I reached for the commission then, and would have crumpled it had he not gripped my wrist. 'Do that,' he said calmly, 'and I'm through with you. For good.' I lay a while, holding the thing against my chest, and it seemed the pain eased fractionally. I watched the square of sunlight move across the old board floor; and already a breeze seemed to blow, fresher than any that troubled the streets of Rome.

It was a fortnight before Marcus pronounced me fit enough to take the road. I had a long journey ahead of me. My posting was to Burdigala, in south-west Gaul; I could find out little about my unit except that they were a mixed Burgundian cohort first raised by Theodosius, seconded by Stilicho for

garrison duty as the bodyguard of one Vidimer, Duke of
Aquitania. Petronius had done his work well; he had succeeded
in interposing the bulk of the Empire between myself and his
erring daughter. Nothing could have suited me better; for
during my illness I had undergone a curious and total change of
heart. The place Julia had held in my thoughts for so long was
empty; try as I might, I could recapture nothing of the feelings
that had obsessed me. As I saw her now, in my mind, I realised
she was not even pretty. I saw a skull too shallow-backed, eyes
too arrogant and cold, a nose too full for beauty; a pouting
mouth that with time would grow sag-lines of discontent, hands
that would turn to mere painted claws. I was free of her, and
free, too, of Rome. From Hispania the city had seemed a
shimmering vision, all that was white and fair. Now, when I
think of Rome, I hear the yelling of the Circus, the endless
echoing thunder of waggon wheels. I see the slaughter pens,
offal heaps mantled with flies, the vomit drying on the wineshop
walls. It's many years now since I left the Mistress of the
World and I know beyond any doubt that I shall never return.

It was a bright, clear morning when we set out. My uncle
came with us to the city wall. Evidently he deemed my departure
a great occasion, for he had put himself to the unheard-of
expense of hiring a chair. Its curtains were drawn wide; within,
Lucullus sat bolt upright with a face of frozen resignation, for
all the world like some general or Emperor suffering an unwanted Triumph. Horses were waiting at the gates; I mounted,
still self-conscious in my new, stiff uniform, expecting him to
dismiss me with a lordly wave. Instead he called me to him; and
I saw instantly by the exaggerated bobbing and rolling of his
head that he had something of more than usual moment to
impart. He stammered and blushed, turning as furiously
crimson as a girl; finally he reached jerkily to grab my hand.
'Boy,' he said, 'you've worked well for me, I'll grant you that;
and I won't have it said that a Paullus ever showed ingratitude.
Here . . .' He dragged a heavy bag from behind him, thrust it
up to me. I started to protest, but he cut me short with such a
furious outburst of swearing that the standers-by at the gate
jerked round in alarm. Another bellow, and the litter, its
bearers equally startled, swung round, headed at a fast trot
back towards the city. The last I saw of my uncle was his
suffused face, ginger wig awry, thrust between the curtains.

He was shouting something incomprehensible while tears—of sorrow at our parting, or grief at his own generosity—welled from his eyes, coursed down his improbable cheeks. Then he and his equipage were gone, lost in the bustle and grinding confusion of Rome.

Four years had passed since I last trod the Via Aurelia. It seemed like half a lifetime. We rode swiftly; the journey, that before had taken so long, was over too soon. We crossed the Cottian Hills, descended to the broad plains beyond; and there came a day when we once more saw, marching on the horizon, the mountains that fringe Hispania. There the ways parted, the Old Road branching off to the left; and there we rested for the night, at an Imperial posting station. In the morning we divided up our gear; for Marcus, who had travelled with me so faithfully, would follow me no longer. He was old, he said, and tired, and if my father would take him back he proposed to spend his remaining years in the place he had grown to think of as his home. I made him promise to visit my mother's tomb, and gave him a purse of golden solidi from my uncle's gift; the price of the horses we had loaned, with interest. I would have written to my father, but when it came to the point I could find no words. Instead I sent a message: that one day I hoped to return, but had not as yet justified myself either in my own eyes or the eyes of the world. Marcus, who had never been one for ceremony, nodded curtly, then reached across to grip my hand. '*Vale*, Sergius,' he said. 'Remember what you've been taught and you'll come to no real harm. Be honest to your superiors, and straightforward with your men; and may the Gods protect you. I hope I shall see you again before I die.' He turned his horse then, rode off without a backward glance; the last I saw of him was his straight back above the cloud of dust raised by the animal's hooves.

I pushed on swiftly, oppressed by the desolation that had fallen on me at finding myself for the first time truly alone, but impelled also by a deeper unease. Beside me, mocking the sunlight, rode a shadow that had followed me from Rome. I had thought myself free of Julia, but it was not so. The wound in my palm had healed, leaving a smooth deep scar like a mark of crucifixion; while it lasted I would remember that the uniform I wore, the horse I rode, the orders and commission in my belt, were no more than crumbs flung from Gratianus' table at the whim of a whining girl. The gold in my panniers I had maybe earned, but none of the rest. I had hidden the thought from my uncle, and from Marcus; but at the bottom of my heart smouldered a black and Celtic rage.

As I neared my destination I tried to put the business from me. The past, I told myself, was the past; over and done with, and best forgotten. I made myself take an interest in the country through which I rode. I was struck by the remarkable recovery that Gaul had made. There was little sign of the disorder that had appalled me; everywhere now seemed to be neat crops and fields, clean and prosperous-looking red-roofed villages. The way stations for the most part had been rebuilt and restaffed; in the whole of my journey I had no trouble with remounts. It seemed that Eusebius had by no means acquitted himself badly in his brief time of glory. Only where his armies had passed did I come on signs of devastation: empty settlements and farms, gangs of ragged-looking refugees. I steered well clear of them, reaching Aquitania without incident. More days of hard riding brought me to the coast, and Burdigala. I arrived in the town towards evening, made enquiries for the headquarters of the Loyal Arcadians. Shortly afterwards I presented myself at the Praetorium of my commanding officer, Vidimerius.

It was obvious at once that the Duke, in the manner of conquerors the world over, lived in considerable style. He had annexed a town house for the use of himself and his staff; I was conducted to it by a towering, heavily armed Burgundian. Orders were barked; I was given into the charge of two equally vast guards, who scanned my papers with a great show of efficiency. Perhaps they understood them; I doubt it, unless they had the ability to read Latin upside down. After more shouting and clashing of heels I was finally admitted to the great man's presence. My first impressions were scarcely encouraging. Vidimerius was by no means as tall as his guards, but what he lacked in height he made up in girth. He was shaped like a barrel, immensely broad in the chest and with thick, muscular legs and arms. A black beard, peppered with grey, covered most of his face; above it small, bleary eyes studied me with no particular affection. The table at which he sat was laden with food; I saw a vast dish of oysters, platters of crab and what looked like mullet, an imposing array of wine. He studied the papers set beside him, painfully deciphering the sentences, before scowling up at me again. He pushed his chair back and favoured me with a long, rich belch. 'Just what the Hell,' he growled, in thickly accented Latin, 'am I supposed to do with you?'

I was considerably taken aback. I answered, sharply, that as far as I was aware I had been posted to command a cohort. He cut me short with a bull-like roar. 'You can forget that bosh,' he said. He slammed the table. The wine-cups jarred and shook. 'Down here,' he said, 'you command nothing. My men don't go much on Romans; they don't speak your language and they don't like your ways. I'm their officer. Me. Not you. As far as they're concerned, and me as well, you're just a flashy ornament. Ever been in action?'

I admitted, stiffly, that I had not. He eyed me, then bellowed again. A man came scurrying from an inner room. 'This,' said the Duke, wagging a thick forefinger in my general direction, 'is the latest asset foisted on us by a grateful State. Find him some quarters.' Then to me, 'Report back here in an hour. Minus that week's growth of beard, and minus the filth on your boots. You look as if you'd been reared in a pig-run.' He finished at a shout, and readdressed himself instantly to his meal.

I was shown to a cubicle at the far end of the villa. The little room was already piled with gear. I turned to the aide, but he merely shrugged. 'Didn't know when to expect you,' he said. 'Haven't got your quarters cleared out. You'll have to share for a night or two.' He indicated an unused bunk, morosely, and left me.

I sat on the edge of it, and rubbed my face. I was oppressively conscious of the strangeness of my surroundings. Round me were unfamiliar sounds, unfamiliar smells. From somewhere close at hand came a burst of laughter, followed by voices raised in Vidimerius' guttural tongue. I stared round; and suddenly the absurdity of the whole situation hit me. Here was I, totally inexperienced in any aspect of war, placed in nominal charge of five hundred men, whose language I couldn't even speak. I didn't feel like a commanding officer, I certainly didn't look like one; it was small wonder I hadn't been treated as one. I put my face in my hands and swore. I'd already lost the first round; but that, I suppose, was likewise scarcely to be wondered at. After all I'd just spent the last two years proving conclusively I wasn't fit to be left in charge of myself.

I was roused by somebody gripping my shoulder. I looked up, startled. Leaning over me was a thin-faced, seamed-skinned man of forty or more. He was casually dressed in tunic and sandals; his eyes, bright brown and half lost in networks of wrinkles, twinkled down at me. 'Flavius Ruricius,' he said. 'Cornicularis, Loyal Arcadians. Welcome to the madhouse, sir; I hear you actually have a civilised tongue in your head.'

The relief of hearing decent Latin again was considerable; I'd virtually decided I would have to go through my military career as a deaf-mute. It was more than offset, though, by annoyance that he had caught me so off my guard. I rose, setting my mouth. 'Sergius Paullus,' I said. 'Tribune . . . I understand my quarters are not prepared.'

'No, sir,' he said. 'Sorry about that. Fix it myself, in the morning. Can I give you a hand with your gear?'

'Thank you,' I said, 'I can manage.'

He squatted on the bunk, undeterred. 'Seen old Rumbleguts yet?' he asked. 'Don't let him put you off. He's been throwing his weight about in all directions for a fortnight now. He didn't want any more Latin-speaking staff people; he's been trying to make this an all-German unit for years. He'd whittled us down to two; now as far as he's concerned the Praefecture's thwarted him again. Had a good journey, sir?'

He was altogether too friendly for my taste, but I couldn't very well throw him out. After all, it was his billet. I changed, hurriedly, and made an attempt at shaving. While I scraped away, with lukewarm water and a blunt razor, he expanded his views on our commanding officer. 'He's a damned good soldier,' he said, 'despite appearances. Not too bad to get on with, as Burgundians go. They're a funny-tempered mob at the best of times, sir. . . .' He also told me a little of the Arcadians' history. As a nearly new unit they had taken no part in the warfare that had attended the elevation of Eusebius. When the usurper's army marched east to its final encounter the cohort had stood and jeered at their less fortunate comrades-in-arms. It would have gone badly for them if the decision of the Frigidus had been reversed; as it was they had received a commendation from Theodosius himself, and the Provincial Praefecture had made up their not inconsiderable arrears of pay. 'Which suited everybody nicely,' said Ruricius sardonically. 'They're a lazy lot of sods when it comes down to it. They'd sooner sit round all day on their arses, milking the countryside round about and steadily increasing the proportion of Gallo-German bastards....'

By the time his exposition was over I was ready to present myself to Vidimerius once more. 'He won't keep you long,' said Ruricius optimistically. 'I'll wait on for you, if you like. Show you a bit of the town.'

Dinner seemed to have put the Duke into a more expansive frame of mind. He sprawled comfortably on a couch, girdle loosened round his ample belly. When I appeared he waved me to a place beside him. 'Here,' he said, 'help yourself to wine. Good stuff. Have it sent down from Germania.' He snorted at my look of surprise. 'We've been growing vines since long before your time,' he said. 'The sooner you people get rid of the idea that civilisation stops at the Rhine, the better. . . .' He heaved himself up, grunting, and poured another libation to his homeland. 'You're serving with a German unit now,' he went on. 'It isn't a question of us coming up to your standards; it's for you to meet ours. Just get that through your head, right from the start.' He drank, noisily. I said nothing; it was obviously best to let him get his introductory lecture off his chest. He rambled on in similar vein for some time before condescending to enlighten me as to the military state of his fragment of Gaul. 'You'll find it pretty quiet in the main,' he said. 'And with that bloody fool Arbogast out of the way let's hope it stays like it. What trouble we've had's mostly been from Bacaudae; but even that's dying off now things are settling down again.'

It was the first time I had heard the word in use. It comes from the Celtic; it's employed fairly loosely to describe any band of homeless or marauding peasants. I remembered when he spoke that some years before serious risings in Armorica had compelled the attentions of an entire Legion. I asked him, with suitable deference, whether there was a chance of similar trouble in Aquitania. He shrugged indifferently. 'There's always an element of risk in a Province like this,' he said. 'Nothing serious at the moment, though. As I said, things are settling down. You only get Bacaudae in any number when the big estates start feeling the squeeze. Most of the poor bastards are starving anyway, it's a shame to knock 'em over. We were called out a couple of years back. About the time Arbogast started making an idiot of himself. Big place a few miles north of here. People couldn't keep going, sold out to the government.

Had to turn 'em off the land at swordpoint. Then they tried to crawl back. All they knew.' He shrugged. 'That's progress for you. . . .'

Ruricius was waiting for me, when I was finally dismissed. Somewhat, I think, to his annoyance, I insisted on first making a circuit of the walls. The endless fighting in Gaul had left its mark on Burdigala, as it had on most of the larger towns. The defences, though in good repair, enclosed a mere fragment of the former extent of the city. Beyond were ruined streets and buildings, dotted with the fires of peasants and squatters. What was left of the town, however, still seemed to me an improvement on Rome. The thoroughfares were broad, well paved and tolerably clean; the place exuded an air of prosperity and peace. Burgundian dress and manners were much in evidence; the Arcadians had obviously made a considerable impression on the local population during their stay. Everywhere I saw wide-bottomed trousers, sleeveless gaudy-patterned jerkins, hair worn long and flowing in imitation of the Germans. Most of the more exotic styles were sported by youngsters in their teens; these, I was interested to hear, were students from the University, which still attracted scholars from many parts of the West. 'Not that I ever fancied ruining my sight with study,' said Ruricius, leering at a stout, matronly woman being borne past in a chair. 'It's the wineshops for me every time. Eh, sir . . .?'

He was a lugubrious character; an incurable chatterer, wholly devoid of ambition. He had entered the Army merely because his father had served before him; like Marcus, he had had no choice. In five years, he told me, he could start looking round for a cabbage patch and a wife; his only fear was that he might find it difficult to secure his release. In these days, with the shortage of manpower in every department of the Army, long-service veterans were frequently refused demobilisation; and the Frigidus had been costly in lives, the Western defence forces were in poorer shape than they had been for years. 'Bad business, that,' said Ruricius, shaking his head. 'Thank God we never got tangled up in it. Saw it coming a mile off, I did. Really bad affair. . . .'

He persuaded me finally into a tavern, where we ran to earth Tonantius, the other occupant of the billet. He turned out to be an Hispanian himself, from Corduba. I spent a pleasant hour swapping local reminiscences while Ruricius listened gloomily, from time to time hopefully rattling a pair of dice. Eventually, when I headed back, I found I couldn't sleep. The streets of Burdigala were quiet; only the calling of the watch, and the bark of a stray dog, disturbed the silence. I missed the rumble and creak of wheels to which I had grown accustomed; the sky was lightening before I closed my eyes.

Later that morning I had my first real glimpse of Burgundians *en masse*. Ruricius took me on a tour of the horse lines and billets. I don't know quite what I was expecting but I know I was agreeably surprised. Everywhere I saw order and neatness, clean-scrubbed walls and floors. Even the master farrier's quarters had more of the air of an office than a workshop, with equipment disposed carefully along the walls, each tool hanging from its own precisely spaced hook. The men themselves paid little attention to me. 'Leave them alone, sir,' said Ruricius sagely, 'and they won't bother you. . . .'

Most of the troopers were big, strapping fellows. I am tall, but some of them topped me by half a head. Popular legend has it they all stand at least seven feet high. That isn't correct, but the other tale that's told about them, that they smear their long hair with butter, certainly is. Why they do it I never really established. What's odd is that in other respects they can be as fastidious as any Roman. I've never seen a German willingly drink stagnant water, for instance, or take meat that's been hung longer than a day; but none of them seem in the slightest put out by the smell of old grease. The rancidness gets in their clothes, clings to any room they enter; in fact give me a favouring wind and I'd pick out a troop of Burgundians from at least half a mile away.

In a way I suppose they're a childish race. They love bright colours, both in their clothing and their surroundings. I became quite hardened to the sight of hulking cavalrymen walking round jangling with necklaces and rings, their long hair tied back like the tails of their own horses. They loved ceremonial too, for its own sake. I found they tended to obey anybody, whether he had legal authority over them or not, provided he made a big enough noise. It was a trait of which Vidimerius took full advantage; they seemed positively to enjoy his tongue-lashings, standing stiffly to attention while he marched round them turning the air blue with his cursing. Yet on some matters they were oddly sensitive. They were very aware of tribal links and ancestries and quick to resent affront, real or imaginary, either to themselves or their clan. Many of them carried detailed genealogies in their heads and could trace back their family history generation after generation to this or that heroic forbear. They were slow-thinking, and unless stirred to anger slow-moving; but once roused they were awe-inspiring fighters. They seemed to have no fear of death and would fling themselves into situations where more sophisticated troops would quail; it was both their strength and their weakness. Vidimerius held them not so much by threats as by the sheer force of loyalty. Each man, I found, had taken a personal oath to serve him; and their word, once given, was their bond. In fact I came to realise they embodied instinctively many of the traits lauded as the essence of the Roman character, with greater justification than we can claim ourselves. I once remarked to the Duke in an unguarded moment that given an army of Burgundians we could hold the world for ever, but it unleashed such a tirade about the perfidy of the Roman State that I was glad to make my escape, and never mentioned the touchy subject again.

On my way back to the Praetorium I questioned Ruricius about their religion. He it seemed neither knew nor cared. Tonantius was afterwards more enlightening. 'A few of them are Christians,' he said, 'but not as many as with some of the other nations. As yet it's only a smattering, and what converts there are are usually bashful about it. Most of the rest still worship old tribal Gods. It's difficult to talk to them and they can get pretty awkward about it, but from what I can make out most of them believe in creatures called the Vanir. They're to do with growth and fertility and tilling the fields; that sort of thing. Then there's the Aesir; they're a higher order that can take on human shape. They have a Divine Triad too: Wotan, Ziu and Donar. Wotan is the lord of storms; he kindles battle-lust in their hearts. Donar gives strength to their sword-arms, and Ziu is the Judge who decides whether or not they're to fall. That's why they're such good soldiers: they believe if they do die fighting, Ziu will take them up and give them a seat in his banqueting hall. Sort of a permanent carouse, with drink and women and all the rest thrown in.'

I thanked him, and complimented him on his detailed knowledge. He looked at me a bit oddly, and muttered something about not being able to avoid picking up bits here and there. I wondered at the time, as I've wondered since, how many Romans are at heart like Tonantius and myself. We're brought up to think nothing of the *bar-bar,* the people outside the Empire who can't even make noises like sensible men; then we meet a few of them and start to have doubts, but we never admit them because it isn't done. Well, so much for fine ideals; they're dead and gone now for the most part, along with a lot of other things.

As events turned out I wasn't allowed to study either Burdigala or my new companions for very long. A few days after I arrived Vidimerius ordered me to mount my first patrol.

Chapter seven

He had had reports, he said, of Bacaudae in the neighbourhood; one of those wandering bands of brigands he'd mentioned to me the night I first arrived. Now seemed as good a time as any to try out my military prowess; I was to take a detachment of a hundred men and sweep the country two days' ride to the north. Maps were produced; Vidimerius jabbed at the parchment with a thick forefinger, indicating last sightings of the enemy and their probable present position. Two villages had been molested, and a man killed very unpleasantly. There was, the Duke suggested, a secondary reason for the expedition. Despite Burdigala's apparent prosperity the town was wholly dependent on grain levies from the surrounding districts; many of the taxes were late in arriving and tended to be of poor quality. Vidimerius, like any good general, never lost an opportunity of supplementing his stores; so if I could see my way clear to procuring the odd couple of cartloads he would be indebted to me. I left with the unspoken understanding that it would be better not to come back than to return without supplies.

I had no clear idea how to set about the job. The Bacaudae, if they existed, had by this time in all probability fled; while wringing grain from overtaxed peasants didn't strike me as part of a soldier's duties. Complaints to the Diocesan Praefecture could lead to a reprimand or worse; and I was equally certain that in the event of trouble Vidimerius would airily disclaim all knowledge of the activities of underlings, especially underlings who couldn't speak his language. All told, I was fairly caught; and Ruricius, when he came to hear of the undertaking, gave no help at all. 'Well, well,' he said, 'it seems he's determined to have his knife into you, sir. What have you been doing to upset him? I've been stuck with the odd job like this myself; but he's usually left the nasty part to his own people, or let me palm him off with excuses. Seems you stand a fair chance of getting your throat cut, sir, one way or another. . . .'

I glared at him. He was either grossly insubordinate or stupid; and I didn't think he was stupid.

None the less the words were hardly conducive to optimism. Still less was I comforted by the sight of the detail I was supposed to lead. They lined up stony-faced, each man staring blankly in front of him. I had managed to secure the services of an interpreter, but he only spoke Celtic and Latin; how I was to control this mob was beyond me. At least they understood the order to walk; we set out from Burdigala in fairly good order but I had an unpleasant feeling that were we to run into trouble they would be better at extricating themselves than rallying to their officer. As it happened the first two days were uneventful. I rode due north, following the coast. On the third morning I began my eastward sweep; when dusk came down I had seen no sign either of the enemy or of any human habitation. This part of Aquitania, though fertile enough, seemed to have suffered more than most in the wars that had swept the Province, and had not as yet been fully resettled. I camped that night uneasily aware that our own rations were running short. Next morning the Arcadians still went about their duties poker-faced; it was obvious it was I, not they, who was on trial. I set my own face in what I hoped was an expressionless mask, and rode on. By mid-afternoon the situation, from being farcical, had grown desperate; I was at my wits' end when we came in sight of the stockade and round-topped huts of a village.

It was a biggish place, covering more ground than I realised at first; but as we came closer my hopes fell once more. It was a wretched little township. Dogs and children skirmished in the dust; at a few hut doors sullen-faced men stood staring, as if daring us to come any closer. There was an air of dilapidation and decay; whatever we might win from the place in the way of stores would barely be worth the effort of carrying away. However, an attempt had to be made. I rode in through the stockade gate, the Burgundians bunching behind me. The headman of the village, or an insanitary, wizened-looking creature I took to be the headman, was shoved forward. I regarded him sternly, and called the interpreter. 'Ask him,' I said, 'what is the name of this place.'

I was rewarded by three or four guttural syllables that told me precisely nothing. I eased my helmet back and wiped my forehead. 'Ask him,' I said, 'if the robbers we seek have passed this way.' The words sounded ridiculous even to me.

The second question was more productive. Both the chief and his subjects burst into voluble abuse; a dozen arms pointed in as many different directions. We were, I took it, supposed to gallop wildly off in pursuit of an imaginary foe. I silenced the din with an upraised hand and tried again. 'Ask the chief,' I said, 'where he stores his grain, and what he might have to spare. Tell him I will write a paper remitting the proper amount at his next tax-giving; tell him also how much we need.'

How well the interpreter made himself understood I couldn't say. This time the whole village burst into uproar; fists were shaken, staffs waved in our faces. The interpreter turned to me. 'No grain,' he said unnecessarily. He added a variety of suggestions, some I suspect of his own invention, as to where we could go for our stores. Any of them could, I suppose, have been construed as treason or rebellion; but I hadn't the heart to set about hanging insolvent peasants. If I had failed, then I had failed; at least I had done my best, though that was small comfort. I turned my horse, stepping back through the dried mud with its circles of old post-holes, to the gate.

Something stopped me. I reined again, thoughtfully. I knew nothing of the ways of these people, but I doubted if their habits were much different from those of the peasants of Hispania. They nearly all dig storage pits to hoard their grain. They line them with clay, and they serve well enough for a season or two; then the damp gets to them, mildew sets in and the simplest answer is to scrap them and dig some more. In that way whole areas get covered with old pit mouths, but I had never yet known peasants to move their huts. I glanced round the compound. There were the storage pits, sure enough; a dozen of them, some still covered by rough gables of straw thatch supported on poles. All were open; all palpably empty. I looked back towards the entrance. There, equally definitely, were the faint annular marks where the huts had previously stood.

I rode back.

'Ask the chief,' I said carefully, 'exactly how he stores his grain.'

The answer, as far as I could make it out, was as I had expected.

I nodded. 'And how long does it take for a pit to sour?'

More voluble explanations.

'Very good,' I said. 'Now ask him, if you will, how many seasons for a hut to stale?'

Before the question was out I saw their faces alter. Hands flew beneath ragged tunics, emerged brandishing daggers. One man ran at me, fumbling to draw an ancient sword. I yelled 'Arcadians...' and rode into him, not wanting to cut him down. He went over handsomely, crouched cowering from the hooves. I heard the screech of drawn steel behind me, a thud and a groan. I wheeled, but the thing was over almost as soon as it had begun. My Germans had deployed neatly, in a half-moon cutting the villagers off from the gate. One man lay doubled up, the grass reddening beneath him; the rest were already flinging down their arms. Faced with the glittering ring of broadswords, there was nothing else to be done. It had been a stupid attack, but they were a stupid, brutal-looking mob. I beckoned to the nearest of the Burgundians. 'Open the walls of a few of these huts,' I told him. 'You'll find as much grain as you want, buried under the floors.'

He must have understood more Latin than he was prepared to admit, for he set to with a will. The others soon caught on; the prisoners were herded to one side and by nightfall we had unearthed more than enough for our needs. I suppose it was robbery, but necessity had already begun to harden me. I called the headman to me again. He came, flanked this time by a pair of burly guards. 'Tell him,' I said, 'he has been guilty of tax evasion, also armed rebellion, the penalty for which, were I to arrest him, would certainly be death.'

He set up a wailing as the words were translated; the Germans thumped him unfeelingly, and he subsided. I stared down at him from the horse. 'Tell him,' I continued, 'that despite this I will take no action, believing his behaviour to be the outcome of stupidity and greed. Also that in spite of his dishonesty I will still give him a note for one half of the amount we have taken from him. The rest I levy as a fine; I've left ample to last the village till next harvest. For his part he will provide a waggon, if these folk own such a thing, and a man to drive it. It may return after it has delivered the grain to Burdigala. Ask him if he considers this just.'

The protestations that ensued left me a little more cheerful; it seemed my victims considered themselves well reprieved. To make my tiny triumph complete, at first light next day we ran into a straggling band of men answering vaguely the description of the people we sought. What followed could scarcely be termed a fight; three or four made a token resistance, the rest were run down as they bolted from the horses. We rode into Burdigala two days later with the fruits of victory; the face of Vidimerius, when he waddled out and spotted the loaded cart, was a memory I treasured for months.

He called me to his quarters that evening to give an account of myself. I explained as concisely as I could how I had come by my windfall. He seemed pleased enough, until I mentioned the remission of the taxes. 'That was a thing you had no right to do,' he said moodily. 'Why couldn't you just collar the bloody lot and be done with it? They'd have been just as satisfied to get away with their skins. A chit from the Army'll carry no weight at all when it comes round to assessment time; we're not supposed to mess with civilian affairs anyway. All it needs is some officious little imp in the Praefecture writing to Mediolanum that we've been currying favour with the natives or inciting rebellion and they'll be looking for somebody's head to put on a dish. Mine, not yours. Though I've no doubt,' he finished darkly, 'we could come to a compromise, as far as that's concerned.'

I answered, curtly, that I had not wished to be unfair; and he burst into a rumble of laughter. 'Unfair?' he said. 'You'll be telling me next the Empire runs on justice and brotherly love. Act your age.' I was silent; and he regarded me curiously. 'You're not one of those chanting, branch-waving damn Christians, are you?' he said. I gathered from his expression that as far as he was concerned that would be the final straw.

I answered him as honestly as I could. 'I was brought up to believe in Christ,' I said, 'but I'm no longer sure.'

He grunted, reaching for the wine. 'Take my tip,' he said. 'Believe in nothing. Except yourself. Get yourself in a hole and no God's going to come swooping out of the sky to get you out of it. That much I've found out. . . .'

I had many occasions, later, to remember his words.

For some weeks after that I was kept busy in and about the Praetorium. What records the regiment had managed to maintain were in a hopeless muddle, as I found when I tried to enter my contribution of grain. Here was one field at least in which a Roman could score over any Burgundian. A room had been prepared for me; I had the books carried to it and applied myself, vigorously at first, to clearing up the confusion. I had inherited some of my father's passion for tidiness and order, and my four years with my uncle had left me with a good grasp of account-keeping if nothing else. I made little progress; and it dawned on me by degrees that nobody, least of all Vidimerius, wanted the books put straight. On paper there were five hundred Loyal Arcadians, but the unit had been under strength since its formation, and the Duke was quite content to have it stay that way. He drew allowances, and pay when it arrived, for a full complement, calmly pocketing the difference. He treated the men fairly and generously enough; what remained he considered spoils of war. That and other devices principally supported his standard of living; some of his evasions were breathtaking, while any queries that did happen to arise were, of course, explained away as the results of language difficulties. I none the less persisted, chasing imaginary totals grimly round and round dog-eared ledgers, till Vidimerius lost patience. 'Much as I admire your industry and learning,' he said, with an access of sarcasm I wouldn't have believed possible in a German, 'I would still be grateful if you kept your fingers out of what doesn't bloody well concern you. . . .' Afterwards he explained his philosophy to me. It had at least the merit of simplicity. 'Everybody cheats everybody else,' he said kindly, as if to a recalcitrant child. 'The farmers and tied men cheat the landlords, the landlords cheat the tax-collectors. The little landlords, that is; the rich ones take care not to pay any tax at all. The collectors cheat the Praefecture; the Praefecture cheats the government; and the government are the biggest damned robbers of the lot. Now one honest man, in the middle of all that, is going to stick out like a sore thumb; he's going to be investigated as an oddity by everyone in the Province, which is just what I don't want to happen to me. . . .' After that there was nothing more to be said, or done.

Life settled into a routine and I for one found time beginning to hang heavily on my hands. Patrols were infrequent and generally uneventful; and in any case Vidimerius was usually content to send his men out in charge of one of their own officers. What he had warned me of was largely true; I was a supernumary, and my days were very much my own. The Burgundians, as Ruricius had pointed out, kept themselves to themselves and apart from the odd grunted word had no real time for any Roman; and I had even less inclination to seek the company of my fellows. Tonantius was pleasant enough when I had to deal with him, but apart from the accident of our origins we had nothing really in common, while Ruricius I have already described. He could be amusing company for an hour over a jar of wine, but his limited range of interests soon began to pall. Mostly, his talk was of women; those he had laid, those he had yet, as he put it, to get his leg across. Occasionally he tried to rouse me to a similar enthusiasm, but the fire that had burned so fiercely in Rome seemed to have reduced itself to ash and embers. If I couldn't understand Ruricius, he certainly couldn't understand me. I overheard him once confiding to the Hispanian that I was an odd sort of character, either stuck-up from birth or as queer as a coot; after that I avoided him more than ever.

Under the circumstances it was probably natural that I should gravitate to the University. It possessed an extremely fine library; I presented myself there one afternoon and asked for the curator. There was some delay before he appeared; my arrival had caused a considerable stir. Once he had recovered from his astonishment at the spectacle of a literate soldier he was more than helpful. I told him I wished to study the known history of Britannia, my mother's Province; it was a project I'd often considered in Rome, and always deferred. He hurried off immediately, returning with an armful of books. 'This is all we have available at the moment,' he said, 'though there is some interesting material I might be able to lay my hands on for you if you so desired. It doesn't seem to be a country that's ever appealed much as a subject for authorship, though it's belonged to us now for nearly four hundred years.'

I told him there was ample to start on; he showed me to a side room where I could work undisturbed, and left me. I sat for a few moments, hands spread flat on the table, eyes closed. The scents of oil, ink and leather that pervaded the place reminded me vividly of my father's study in Italica; it took little effort of imagination to convince myself I was back there, that the last four hectic years had never intervened. Any moment now I would hear Ursula's quick, scurrying step pass the door, or the soft swishing of my mother's robe. I opened my eyes, shook my head and settled down to work. I read till nightfall; much still lay untouched when the library closed for the day. I left considerably refreshed; for the moment at least my problems were solved.

Shortly afterwards, the even tenor of life in Burdigala was disturbed by news of the death of Theodosius and the subsequent elevation of his son. Some there were who found the prospect of a boy Augustus alarming, but Vidimer was not one of them. 'Stilicho's been made regent,' he said. 'And he's a good man, although he's a bloody Vandal. Marched with him once, just after we were formed. Good fighting man.' For Vidimerius of all folk to make such an admission was remarkable, for there was little love lost between any of the Germanic tribes. It was the one thing, as the Duke himself had pointed out, that had prevented them from uniting years back in the time of anarchy and sweeping us completely off the board. 'What's more to the point,' said my commanding officer, running his tongue reflectively round his lips, 'is what we stand to get out of it. Pay's far enough behind again already, the Gods know. . . . Should be worth a few solidi apiece at the very least.' He fixed me with a sudden beady eye. 'You'll see a bit of fun then,' he said. 'You mark my words. . . .'

In the event, Honorius paid a handsome donative. Vidimer at least made sure of his share of the prize; a full half of the strength of the Arcadians was despatched to the borders of Aquitania to escort the bullion waggon home. The excitement the Duke had prophesied followed soon after.

The regiment paraded at first light the day after the money arrived; Vidimerius presided grandly over the occasion, seated at a trestle table heaped enticingly with dully glinting stacks of coins. To either side a huge Burgundian, sword unsheathed, ensured the equitable distribution of the bounty. The Duke seemed already somewhat the worse for wear; his little piggy eyes, outlined now in festive red, watched suspiciously as each man stepped up, saluted smartly, gathered his money and withdrew to the ranks. Our shares were left till last; Vidimerius gave them to us personally, accompanying each with an unpleasant leer and an admonition to use the money well. I had little doubt what Ruricius at least would do with his; mine went to swell my personal savings, which I had already lodged, through the good offices of my librarian friend, in the University safe.

I would have returned to my books, but I found the library, in common with the rest of the public buildings, closed down. There was good cause. I would not have believed that four hundred dour and massive Germans could have been so rapidly transformed had I not the evidence of my own eyes. By midday most of Vidimerius' men were roaring drunk; by nightfall they were like fiends from the Pit. They roamed the streets in droves, chanting and bellowing; it was a brave citizen, or one with pressing business, who showed his nose beyond his own front doors. A few wineshops attempted, unwisely, to put up their shutters. It was to no avail; the Arcadians simply removed them, by the quickest means that came to hand. In time, as establishment after establishment ran dry, the din lessened somewhat, but the streets were still far from safe. One couldn't step far in the blackness without tripping over a snoring German; and there was always the possibility that he might settle the question of rights of way with an axe or sword. At dawn Burdigala wore a beleaguered look, its houses bolted and barred, its streets strewn with debris; what Arcadians remained on their feet tottered about uncertainly, grey in the dim light, like ghosts from their own fabulous Hell. But by midday the uproar was as bad as ever again, for many traders in the vicinity, scenting quick profits, had hastened to load cargoes of wine for despatch to the town. The first consignment never reached the walls; after that some sort of discipline was imposed and things became more orderly, though far from normal. It was amazing to me that nobody was killed; our only casualties were a cavalryman severely bitten by a pig, and Ruricius, who succeeded sometime during the fracas in falling from top to bottom of a flight of stairs. He was lucky to break only his collarbone, and not his neck.

A week or so after peace was finally restored I received an invitation to dinner at the house of the University librarian, Sallustius Patermuthis. I accepted eagerly. The visit would be a welcome break with routine; he was a kindly man, retiring and shy for all his learning, with whom I had already formed a tentative friendship. Frequently on my visits to the library Patermuthis would call me to his office to examine this or that manuscript or map, for my interest in Britannia had stimulated him to research of his own, and he liked nothing better than to impart some titbit of freshly-gleaned information. Also the invitation was a signal honour; the curator was well thought of in Burdigala, and highly influential at the University.

He was a bachelor and lived simply, attended by a minimum of servants. His house adjoined the library building itself; I arrived punctually on the afternoon named, to be met by Patermuthis himself. He ushered me to the baths, where several slaves waited my appearance. I took my time; a couple of hours later my clothes were returned to me and I made my way back, refreshed and relaxed, to the peristyle.

The house was very much bigger than it had appeared from the street, and furnished with a taste and elegance seldom encountered outside Rome. Fountains played softly; I was reminded once more, forcibly, of my father's home. The dining room itself was the biggest chamber in the place. It served double duty as a study; three walls were lined with books, while to one side stood a massive writing desk, its top littered with scrolls and stacks of paper. Once over the threshold I paused involuntarily. The whole floor was taken up by a mosaic, worked in a simple, powerful style I had not seen before. Heads of the Seasons stared from the corners; I saw Spring with her chaplet of green leaves, Winter hooded and stern, clutching the bare branch of a tree. Between were patterns of flowers and birds, while in the centre was a scene from mythology; Venus, attended by handmaidens, rising from the sea. Here the artists had exercised their greatest skill. The hair of the Goddess, corn-coloured and long, flew in a non-existent breeze; her eyes, delicately tilted beneath arching brows, watched up at me coolly. She wore a cloak, fastened at the neck; but it too flowed up and away behind her, leaving her naked. Her body, slender and delicately modelled, glowed against a background of blue-grey and green. One slim foot was poised, in the act of stepping from the water; beside her a girl stretched out her hand, ready to assist her to the shore.

I became belatedly aware of Patermuthis waiting by my side. I apologised, awkwardly, for my rudeness, but he merely laughed, steering me towards the eating alcove. 'It is in fact a most delightful work,' he said. 'The main charm of the house, to my eyes. The Bishop, I'm afraid, scarcely shares my enthusiasm; he has gone so far as to suggest that strategically placed matting might be an asset to the room, but on that point I've so far remained obdurate. Please be seated, Tribune, and make yourself at home.' He took the couch opposite me, and clapped his hands for wine.

I was more than a little surprised to find myself the only guest. Patermuthis understood my look, and hastened to explain. 'I've always been averse to crowds,' he said. 'And a man sitting at meat is, after all, at his most primitive; he craves, and deserves, a certain decent solitude. Had I nine heads, and as many tongues and ears, I would cheerfully fill my couches every night; as it is, I believe if a man is worth inviting to one's home he's worth listening to, and the least one can give him is undivided attention.'

'I think,' I said, 'my father would have agreed with you there, sir.'

He signalled to the wine servant to refill my cup. 'Ah yes,' he said, 'your father. I'd like to hear more of him, and of you. I must confess to a certain curiosity. You've obviously had an excellent upbringing and education; how on earth did you come to finish up, of all things, in the Army?'

The meal was wholesome: meat dishes and fresh-caught sea fish, served with simple, effective sauces. Under the influence of that excellent board I spoke freely, for the first time in years; I described my life in Italica and the events that had taken me to Rome, omitting only the still-painful details of my abortive love affair. Patermuthis heard me out thoughtfully, nodding his head from time to time. 'Well,' he said at length, 'that certainly explains your interest in Britannia; and you appear to have inherited your father's enquiring mind. But if you'll forgive a somewhat personal remark, you seem to have made a grave error somewhere along the line.'

I took a sip of wine. 'How do you mean, sir?' I asked him.

He smiled faintly, and let his eyes drift back to the Venus before answering. 'Let me express myself in a different way,' he said. 'Here you are, your ambition achieved, a Tribune on active service stationed in Burdigala. Are you happy?'

I was quiet for a moment, considering. 'I don't know,' I

said finally. 'Define happiness and perhaps I could answer you.'

He laughed at that, not unkindly. 'Don't split hairs, young man,' he said, 'or chop your Greek logic quite so fine. I asked you a simple question, which seems to me to have an equally simple answer; no. Let's see, to begin with, what Burdigala has to offer in entertainment for a fellow of your type. The wine-shops, of course, where I'm sure that rather unsavoury little fellow officer of yours would be delighted to keep you company; the brothels . . . though it seems to me you're not exactly the sort of man to find forgetfulness in the comforting arms of an whore. Is there anything else?'

'Very little,' I admitted.

'No,' he said. 'Very little, for a soldier.' He looked at me intently, frowning. 'Sergius,' he said, 'it's not too late, if you'll take counsel from an older and perhaps a little wiser man. You have some money of your own, as I know perfectly well; not a great fortune, but more than enough to buy you a release. I hope I haven't offended you by saying you've made a mistake; you see it's a mistake that can be very soon remedied.' He waved a hand in a vague gesture at the shelves of books behind him. 'It seems to me,' he said, 'that you were destined for something much better than an army career. You've had a good education, as I said; and you've a better brain than many it's been my misfortune to instruct over the years. If you wished to take up your studies again, here, you'd find I could be of considerable help to you. You could read law; or perhaps a position could be found for you within the University itself. In time you could even become a tutor; I feel you might make a good one. While as far as the present is concerned, as you've already seen this house is really much too big for one. You could live here, if you so desired; at least until you became established. In a curious way I've come to feel a certain responsibility for you; I urge you at least to think over what I've said, and not give too hasty an answer.'

I sat quiet, frowning down at my wine-cup. Many things passed through my mind. The offer was as generous as it was unexpected; certainly I would never be given a better chance to start again. And much of what Patermuthis had said was uncomfortably accurate. I had seen now, at first hand, just what modern army service was like; and I knew in my heart I could never give myself wholly to my chosen career. But ... I found myself unconsciously rubbing the new scar in my palm. I thought of what my commission had already cost me in mental suffering and pain. It wasn't a thing to be lightly put away. Maybe, too, those distant childhood trumpets rang in my mind again. One day, I knew, the Arcadians would ride out from Burdigala, to battle and glory, maybe death. Could I stand to watch them, one of a faceless crowd again?

I looked down at the Venus. Her eyes still seemed to seek me out where I lay; and I experienced the oddest sensation of confusion and uncertainty. I saw, in a moment of time, that my fate was not yet accomplished. What was it Marcus had said, all those years ago? 'It's as if we were launched on a sea and must sail on till we reach land, wherever it might be. . . .' I had felt myself beyond all love for woman; now I knew with curious sureness that this was not so. In the face of the Anadyomene, human yet divine, I saw perfection. Maybe her counterpart existed, on the broad earth, and breathed at this instant and talked; maybe I just pursued an ideal. Either way I must continue to search, perhaps across the ocean.

I came round with a start. My host was watching me half-humorously. 'Well,' he said, 'I'm sorry my suggestion came as such an unwelcome shock; for a moment you looked positively appalled.' I hurriedly apologised, but he clapped his hands again, and laughed. 'At any rate,' he said, 'let's drink to your future, whatever it might be. And for the present please consider this house as your home. I'll have Hoenus make up a bed in one of the spare rooms; you can use it whenever you choose, or whenever you feel the exigencies of military life about to overcome the scholar in you.'

I drank to that, in gratitude; and when I think back over my time in Burdigala it is always the house of Patermuthis, and the hospitality I found there, that come first to mind. The possibility of my leaving the Army was never discussed again, but his interest in me was unflagging. Through him I met many of the leading citizens of the town. As he had pointed out, he entertained sparingly; but when he did give a dinner I was always assured of a place. He would introduce me, smilingly, as a military historian, and would often attempt to draw me out in discussion with my elders. I committed myself as little as possible on these occasions; for to tell the truth I found I had little more in common with the intelligentsia of the town than with the Arcadians. Authors abounded in Burdigala, but to me each seemed as impossibly tedious as the last. They would entertain the table in turn with their latest odes, applauding each other vigorously; at the end of each rendition Patermuthis would incline his head gravely, murmuring this or that compliment to the reader's style or diction, while I came near to bursting. To me their efforts seemed devoid alike of interest and originality; none of these literary Tribunes, curiales and priests seemed to have the slightest real appreciation of the structure of language, the golden ring of finely chosen words. I mentioned the matter once to Patermuthis. We had been discussing the work of Cicero, of whose writings he possessed a remarkably fine collection; he laid down the book from which he had been reading, shook his head and smiled. 'You've chosen a hard course, I'm afraid, my friend,' he said. 'The way of the poet. You'll find few enough to agree with you, even in Burdigala. Folk flock from half across the Empire to sit here at the feet of masters; and what do they learn? How to cram their work fuller with Classical allusions than a beggar's dog is full of fleas, but precious little else. So that Gallia abound from end to end with pompous little orators shedding panegyrics like trees shed autumn leaves; each one a Cicero, mark you, in his own estimation. And, alas, in the eyes of the bedazzled population.'

He was even more scathing about the attitude of the Gallic nobility to events within the Empire. 'Within living memory,' he said once, 'this Province has been swept by war after disastrous war. We've seen not merely tribes but entire barbarian nations admitted, with their Kings, their customs and their unrest, into the Empire itself. We've seen a complete Roman field army routed at Hadrianopolis by these selfsame savages, and an Augustus killed. We've seen another of these harmless buffoons proclaim his own State, set up his own Emperor and make idiots of the entire administration of the West. Right now there are more barbarians under arms inside our borders than at any time in history; men like your Vidimerius, bound by the most tenuous loyalties to a culture that at best they barely comprehend, each ready, were conditions to alter, to carve out little principalities for themselves and hold them against all comers. You see soldiers, Sergius, and you may well be right; I see the very instruments of chaos.'

I was amazed that Patermuthis, the mildest of men and the last, one might have thought, to concern himself with military affairs, should show passion in such an unlikely cause. I said as much; and he smiled, a little bitterly.

'I'm not so readily moved to anger as I was,' he said, 'but this infuriates me. When Rome made these lands into a Province, four hundred years and more ago, she took over a mass of peasants and warring, petty chieftains; and that, despite our fine clothes and finer words, is just what we've remained. But we look to Rome now for our protection, as if there was some magic in the very name; we forget how she herself was founded, or how she grew. We build no walls; we send her no men; when the enlistment orders go out we whine to the Praefecture that our estates will fall into ruin, and commute levies for gold. While the literary among us, in between playing with our books and dinner parties, bemoan the decay of virtues that were never ours in the first place and blame the results of indolence, greed and blindness on the wrath of the Gods. Right now, as we sit and talk, the northern frontiers are menaced from end to end, by peoples united, for the very first time, by a common terror as well as a common lust for plunder. It's a pressure I can feel, as a nearly physical thing; I feel it with my bones. It appals me, even if it doesn't frighten you. What do these people see, beyond the towers and walls, when they look down to the south? An entire world, vaster than their minds can grasp, richer than all their imaginings, lying like a great plum ripe for picking. And what do we do in the meantime? Put our fingers in our ears, turn our cloaks over our eyes, our arses to the north and pretend we never heard of Germania. Here, I'll show you something....'

He rose, stamped to his desk and unlocked a drawer. From it he took a book that he slung across the room to me. I slipped it from its bright-dyed jacket, wondering. It was entitled *A Treatise on War,* but there was no author's name. I began to read, and was soon engrossed. Here, clearly and concisely, were set out reforms that touched on every aspect of the Empire. Firstly the monarchy was attacked for its lavishness and abuse of public funds. There followed detailed plans for financial reform that included a new and flexible tax structure; levies were to be made in strict accordance with the means of the individual, and major landowners, who at present escaped virtually unscathed, compelled to bear their full share of the burden. These landowners, too, were to be made responsible for the maintenance of a system of frontier forts, and the entire Army reorganised on rational and sensible lines, new regiments replacing the hotch-potch of auxiliary units raised by this Emperor or that and as often as not left to disintegrate slowly in whatever outlandish spot their service had chanced to leave them. I skipped hastily from paragraph to paragraph. Here were no woolly declamations on the virtue of our ancestors, but commonsense, hard-headed solutions to the immediate problems of the day. I saw new war machines described; armoured vehicles in which small detachments of men could successfully engage much greater concentrations of enemy troops; pontoons that could be flung instantly across the deepest river, from which bridgeheads could be established in the face of entrenched opposition; even a warship powered by oxen, which could manœuvre independently of the wind and yet leave its entire crew free to engage the enemy. This last in particular intrigued me. I pored over the illustration of the craft. The animals, I saw, worked a treadmill on the upper deck; from it shafts led to twin wheels projecting over the bulwarks. Oar blades were attached to their rims; it was a simple enough device, but I wondered at the power it might develop. Such a vessel, properly armoured and equipped with a heavy ram, could prove invincible.

I laid the book down finally and looked up. Patermuthis was watching me. 'Who wrote this, sir?' I asked him.

He took the manuscript from me. 'I did,' he said. 'When I was younger and more enthusiastic, and felt the Empire might yet be saved.' He rolled the parchment neatly, slid it back into its jacket. 'Very few have ever seen it,' he said. 'Those who have consider it a literary curiosity but nothing more. Perhaps they're right. Nobody, except you, knows the identity of its author; and I'll charge you, if my friendship means anything at all, to keep that to yourself.'

'But, sir,' I said, 'I've never seen ideas like that before. They should be put to the test. It should be sent to the Emperor.'

'It was,' he said shortly. 'The noble Valentinian was not amused. For my part I've become convinced no single voice will ever now be heard, much less heeded. The Empire will go its own way, regardless of what you or I might wish. Even to acknowledge that thing, now, would bring down more ridicule on my head than I could stand. What, folk would ask, does a librarian know of soldiering? Or an historian of machines of war? No, Sergius, let it be. . . .'

He was my mentor; I could do no other than respect his wishes. He is dead now, and I hope at rest; so his secret can do no harm. As for the book itself, I've wondered since what happened to it. Burned, perhaps, in the wreck of half the world, that wreck it so vividly, and so uselessly, foresaw.

Chapter eight

Between my visits to the home of Patermuthis I practised hard at weapons drill. My life in Rome, I felt, had softened me considerably. I set myself a rigorous training programme, and adhered to it through what remained of the year. I was afraid the new injury to my hand might have once more weakened my grip; certainly I was clumsy at first but I found my old skill rapidly returning. With the short sword I was more than a match for any of the Arcadians, though I was less happy with the long German spatha and never came to terms with their other favourite weapon, the axe. It's a frightful instrument. Swung or hurled, it's capable of inflicting ghastly wounds; and no shield will ever stand against it. The Arcadians were adepts in its use, charging home against the practice stakes which they splintered with fearsome yells. Sheer strength, of course, accounted for a good deal; though they possessed an unerring aim and for all their bulk could be as nimble as cats. One and all prided themselves on their skill and fitness for battle; I was reminded again, watching them in training, what unpleasant adversaries they would make.

Vidimerius, when he came to hear of my preoccupation, expressed his approval in no uncertain terms. Nothing would suffice but that the staff officers set aside a period each day to join me in my practice. It was a development I had not foreseen; it destroyed my remaining stock with Ruricius completely. As ever, the wineshops beckoned him; he seemed to feel puffing and grunting on a parade ground was no part of life for a soldier of Rome. I avoided, as far as was possible, further ruffling his feelings, but I might have spared my efforts. On the evening when Vidimerius came in person to inspect our progress Ruricius threw caution to the winds, attacking me with something very like murder in his eyes. He made, I think, a mistake that was to be made again. I hold a sword as a gourmet holds a wine-cup, delicately, with the little finger raised. But the grip is stronger than it looks. I had the advantage in age, weight and reach; added to which he was in a thoroughly flabby condition, and an indifferent swordsman at the best of times. A prod from the blunted practice weapon doubled him up; I tapped him lightly above the ear, and sidestepped. He measured his length, cursing, while Vidimerius, roaring with delight, made immediate preparations to engage me himself. He was a crafty, dangerous fighter; in the end I barged him as Marcus had once taught and he joined Ruricius on the ground, my blade prodding his throat. We fought twice more. Both bouts were equally decisive, so much so that I began to fear I would suffer later for showing him up in front of his men. I needn't have worried; whatever faults a German may have, ungenerousness is not usually one of them. Vidimerius was highly amused, insisting I should join him at dinner. Over his habitual trencher of oysters he asked me how I had come by my skill. 'I thought you spent your days messing about in libraries,' he said. I answered, straight-faced, that I grubbed it out of my books; he stared at me suspiciously for a while, but said no more.

Overall, my life in Burdigala was extremely pleasant. The summers were warm and long; I spent days in sea and river fishing, or riding in the surrounding countryside. For once the frontiers of the Empire were quiet; dangerously quiet, perhaps. What unrest existed on the borders made no impression on us deep in Gallia.

Winter, and a succession of iron-hard frosts, put an end to such diversions. Christmas passed before the weather broke, with a month or more of deluging rain. The river rose, flooding half the town; the Germans, confined to their barracks, grew sullen and awkward-tempered. Vidimerius took to watching his men anxiously. What was needed, he told me in a sudden access of confidence, was a spring campaign. There had been no opportunity lately for the Arcadians to win either plunder or glory in battle. Under the circumstances it was natural they should become restive; unless their attention was diverted by a little healthy warfare they would almost certainly start to desert, drifting back in twos and threes to their homeland.

The Duke viewed the prospect dismally; he at least had no wish to return to his old way of life. His father, he told me, had been a petty chieftain across the Rhine; many of the Arcadians were descendants of his original war band. He described to me something of tribal life and custom; how able-bodied warriors would gather at the hall of a successful chieftain, whose popularity depended thereafter on the quantity of spoils he gained in war. It was an attitude too deeply ingrained to break; the Arcadians, despite their nominal status as soldiers of the West, still thought largely in such parochial terms, looking to Vidimerius for a steady supply of booty and excitement. A mutiny, he admitted, was by no means out of the question; he warned me to handle the men, when I had to, with the lightest possible rein.

There were other reasons for caution. Under Theodosius—who had borne, among his other honours, the somewhat lugubrious title 'Friend of the Goths, and of Peace'—Germanic tribesmen serving within the Empire had been treated with toleration. Now, with the Emperor gone, the older, deep-rooted distrust of barbarians was once more making itself felt. I was conscious of the altered atmosphere even in Burdigala; folk were beginning to look askance at Vidimerius' men, and avoid them in the streets. The Arcadians, for their part, took to going about in bands, all heavily armed; the massacre after Hadrianopolis had by no means been forgotten. When their pay once more failed to arrive the Duke quietly made up the deficit from his own pocket; he could have offered no more convincing proof of the seriousness of the situation. A few days later he took it on himself to summon the duovirs of the town to his headquarters. Burdigala, he told the shocked curiales, must be prepared to dig into its own pocket; to amuse his men he proposed holding games in true Roman fashion, in honour of the German Trinity. The glad news would be passed to the Arcadians without delay; after that, if Burdigalans valued their skins, the entertainment had better be forthcoming.

As things turned out, the plan was never put into effect. The reprieve came in early April, in the form of an urgent despatch from Mediolanum. Once more the roads of Gallia were passable; in the north campaigning had already begun. During the winter, war bands of Alamanni and Franks, supported by Saxon pirates from the mouth of the Rhine, had succeeded in establishing themselves on the coast of Belgica. Siege had been laid to Remi and Rotomagus, and a strong auxiliary force was being raised to disperse the menace. It would be backed by detachments from Marcus' old Legion, II Italica, and commanded by the Frank Merobaudes. We were ordered to force march with half our parade strength, and rendezvous with the main column at the old Celtic settlement of Parisi.

Vidimerius was converted on the instant to roaring good humour. Orders were bellowed, horns blared; within a remarkably short time decurions and captains of centuries were gathering at the Praetorium. 'Damn this bit about half our strength,' said the Duke scornfully. 'I want you to round up everybody you can lay your hands on. This lot have been on the rampage for weeks, they're bound to be fat. Best chance of pickings we've had since we came down here.'

So much, I thought, for Roman discipline. In the end Ruricius was left behind, to his unconcealed delight, with a handful of the least warlike of the Arcadians, to oversee the defence of Burdigala in our absence. The rest of the cohort paraded a couple of mornings later. Sitting my horse beside the town gates as the column debouched to the north, I thought I had never seen an untidier-looking rabble. At its head marched trumpeters, sounding from time to time the long, harsh-throated horns of which all Germans are inordinately fond. Next came Vidimerius, clad for the occasion in a vast and complicated war helmet, all knobs and rusty iron points. A heavy night at mess had left him in far from the best of tempers; he was bellowing himself crimson in the face in attempts to close the men up into some sort of order. Behind him rode the cavalry on their heavy, shambling horses. No two troopers were dressed alike; each vied with his neighbours in the gaudiness of his attire. Multi-coloured cloaks fluttered and swirled, pennants floated from the heads of lances. Behind again were the infantry, in a straggling mass that occupied the entire width of the road. On their heels lumbered the supply train. Each waggon was piled high with camp-followers, mainly women; it seemed half the whores of Aquitania had embarked on the enterprise, to see at first hand this amusing game of war. At the rear, last but by no means least, trundled a heavy onager, Burdigala's one and only war engine and the apple of Vidimerius' eye. I had never seen it used. Onagers are bulky, unhandy affairs; the tails of these engines are apt to kick violently as the shot is discharged, earning them their name of wild asses. Why the Duke wanted to bring one on the march I couldn't imagine, except that it pleased the specialised sense of humour of Germans occasionally to pelt the heads of captured enemies about the countryside. I saw the contraption clear of the gates, nodded a curt farewell to Ruricius and rode ahead towards the front of the column.

Several days' march, skirting the coast most of the way, brought us to the border of Lugdunensis, where we turned inland for Parisi. The cumbrousness of the train caused innumerable delays; in the end Vidimerius left most of the waggons to follow behind under light guard and we made better time. A few miles from our destination we sighted the standards of Italica. This was considerably better. The column, cohorts of horses and infantry and a mixed bag of Galician and Cantabrian auxiliaries, was marching in good order, point troops flung out in textbook fashion to either side of the road. Merobaudes was leading the van. He and Vidimerius greeted each other boisterously; they were apparently old acquaintances. The Arcadian cavalry closed up ahead of the Legionaries. Our infantry were posted to the rear, after the last Galician contingent but preceding the train; for no German will follow baggage carts. The dispositions, necessary though they were, occupied some little time; it was late afternoon before I rode into the town. I found Parisi scruffy and malodorous; what public buildings existed looked dilapidated in the extreme, while the streets were thick with a litter of straw and ordure. Under the circumstances it was odd to hear strongly inflected Latin being spoken. The place still possessed a town senate of sorts; the curiales, resplendent in togas embellished with vast faded purple stripes, delivered an address of welcome followed by an attempt at a panegyric lauding the noble and peace-loving nation of the Franks. Merobaudes, whose presence had only been made necessary by the depredations of his countrymen, cut the oration short with a curt demand for supplies. Horses and men were quartered in the town; and I for one passed a restless night, turning and tossing on a straw pallet shared with what seemed an entire Legion of minute but voracious bedfellows.

We were joined shortly after first light by a further Pannonian contingent and a dashing troupe of Hispanian horse; after that the forces divided, Merobaudes taking the Italicans and Pannonians on to Belgica, the rest under Vidimerius turning west to the relief of Rotomagus. Most of the talk in the mess was in clanging German; from it I gathered that Remi was already clear of the enemy, who had fled at our approach. My commanding officer was more distressed by the failure of his beloved onager to put in an appearance than by the other difficulties of the march, considerable though they were. The column spoke three or four tongues; I spent an harassed morning dashing forwards and back along the line trying to instil some sort of order. We camped that night twenty miles from Parisi. At dawn despatch riders reached us from the coast with news that units of the British Fleet, riding menacingly a mile or so from shore, had put most of the Saxons to flight; there had been a skirmish, and some ships had been burned. Rotomagus was no longer in danger; the bulk of the invaders had gone north, hoping probably to reach some rendezvous further up the coast. Vidimerius, after some grumbling indecision, turned north himself; the messengers were sent on to appraise Merobaudes of the situation, and the column lumbered in pursuit.

So far we had seen no sign of the enemy; now our scouts reported concentrations of war bands from a dozen or more points ahead. At our speed of march we were unlikely to overtake them; Vidimerius detached the cavalry, both Hispanian and Burgundian. They charged for the horizon in fine style, uttering piercing whoops. It seemed to me a most unsatisfactory decision; the column was left without an effective screen, while the lancers, unsupported by infantry, could find themselves in difficulties if they did happen on the enemy in any strength. The Duke later condescended to explain his reasoning to me. His own men, he said, would not have remained in check much longer once they had scented plunder; if it came to a choice he would rather lose a few by enemy action than by defection, while the weaker the main column appeared the more chance there was that the raiders would be persuaded to make some sort of stand. My opinion must have been reflected in my expression, for he growled disgustedly. 'All your diagrams and bits of paper might look well enough in libraries,' he said, 'but they're no damned good for Germans in the field. . . .' He was right, of course. He knew his people far better than I; I came to realise by degrees that despite their many excellent qualities, trying to shape Burgundians into a conventional Roman force would have been like trying to model a statue from water.

That night we made our first marching camp of the campaign. The Galicians did most of the work; the Germans, touchy as ever on points of honour, considered it beneath their dignity to wield a spade. What tents we had with us were pitched; most of the Arcadians slept in the open air. Night fell with no sign of the cavalry. Vidimerius was unperturbed. The detachments began to straggle in an hour after dark, in no particular order. They brought booty with them, mainly weapons, and a handful of prisoners. It was the first time I had seen barbarians from beyond the Empire at close quarters. They were big men for the most part, with streaming, matted fair hair. Many were injured, some severely, but no attempt had been made to dress the wounds. They had been stripped by their captors; all bore marks of ill-treatment. Vidimerius examined them critically before ordering them to be placed under close guard. A few, he told me, might make useful additions to the Arcadians' ranks; the rest would be killed at a more convenient time, either in combat with wild animals or each other. Appalled though I was, I realised that for a German it was a merciful decision; he was giving the creatures the opportunity of an honourable death, the passport to their own strange Heaven.

The main body of Arcadians clattered in as the first watches were being set. They brought more specific news. A few miles ahead, Frankish war parties had occupied an ancient hill fort. Their numbers were uncertain but seemed to be considerable. The patrol, pursuing a fleeing band of Alamanni, had come in sight of the place at dusk. They reported many fires, ramparts lined with warriors. Exercising prudence for once in their careers, they had withdrawn, leaving half a dozen men on the watch for further activity from the raiders.

Vidimerius was delighted; nothing could have suited his schemes or inclinations better. He called me instantly to attend him. He proposed, he said, to examine the stronghold in person. Horses were saddled; half an hour later a small party rode north.

The night was fine, with a clear, high moon. For most of the journey we followed a metalled road; it stretched ahead dim in the moonlight, striped and barred by the shadows of trees. The ground round about was thickly wooded; bad marching country but ideal for guerilla operations. The Germans at least were unconcerned; they rode steadily, harness clinking and rattling, long hair flying. The Duke seemed lost in thought. He only spoke once, gesturing briefly at the sky. I glanced up, following the direction of his arm. The night was still fine, but a solitary cloudstreak had crawled from the horizon, it seemed across our course. As I watched, the moon was extinguished; the cloud-tip glowed with ragged edges of silver. 'Wotan's daughters,' said Vidimerius grimly. 'Come to choose the slain. An omen of battle, my friend. . . .' He relapsed into silence; I shivered, drawing my cloak more firmly round me. Beneath the trees the air struck suddenly chill.

The enemy camp advertised its presence for some miles. Fires burned on the ramparts, twinkling spots high in the night. As we came nearer to the place we slowed. It lay some halfmile back from the road, crowning the tip of a wooded spur of hills. In front of it the ground was broken and confused, crisscrossed with gullies and overgrown by scrub. To the south, immediately on our right, the forest thrust out a long arm of trees; it lay dense and black in the night, reaching to within a few hundred paces of the road.

Vidimerius rode on thoughtfully, staring up. Originally the embankments of the fort would have been crowned with palisades. All had long since been overthrown, tumbled into the ditches; but the smooth grassy slopes, overgrown here and there with clumps of bushes, still presented a considerable obstacle. I tried to imagine how I would defend such a place, were I its commander. Its weakest points would obviously be the gates, which our scouts had informed us were three in number: one in the middle of the southern face, one each at the western and eastern extremities. The timbers that had closed them would likewise have vanished, but piled brushwood could still provide an obstacle sufficient to slow a charge. Also we would be forced to attack on a narrow front; I would man the flanking earthworks heavily, pick off horses and riders as they struggled to break through. I wondered how many men we could expect to oppose us. From my viewpoint twenty fires were visible. I counted again, carefully. Double that number for the whole circumference, add a few more; and assume the invaders wouldn't give themselves more trouble than was necessary in the matter of gathering fuel. Say, then, ten men to a fire; unless the fires themselves were a bluff, designed to confuse our notions of the raiders' strength. But that could be discounted. They were in strength, and confident; otherwise they would simply have melted away into the woods.

Vidimerius seemed to have been having similar thoughts. He spoke abruptly. 'How many men in there, judging from what you see?'

I said, 'Five hundred,' promptly. He stared at me for a moment, pulling his beard, then grunted non-committally. We walked on again, unchallenged. We were abreast of the fortress before I heard faint sounds of horsemen. Tall shadows materialised from the night: the detachment the Arcadians had left on guard. A muttered conversation ensued, in German. I saw the leader spread his fingers several times. Finally Vidimerius turned to me. 'Five hundred was a good enough guess,' he said. 'We've got 'em in the bag.' He ordered fires to be lit, many fires, to north and south of the fortification. The Franks, if convinced of the presence of a besieging force, might at least hold the position till daylight. An hour passed in the work; then we turned south again. The sky was brightening as we came in sight of our camp. I retired to my tent, to catch an hour's sleep if possible before the march.

It seemed the war horns roused me as soon as I had closed my eyes. I hurried to the Praetorium. Dawn was barely in the sky, but Vidimerius, who apparently never suffered from fatigue, was already shouting orders. Hubbub rose from the cavalry lines; men scurried in all directions, saddling horses, buckling on weapons and armour. I snatched a hurried breakfast of fruit and bread, listening intently while the Duke outlined his plan of attack, mapping out the enemy position with the aid of a stick and a tray of sand. Half an hour later the column swung north on to its line of march.

Vidimerius set a hard pace. Five miles from our objective we were met by Arcadian scouts. A further trickle of Alamanni had come in, they said, during the night. A few had never reached the fortress. They displayed ghastly trophies, grinning broadly. Several attacks had been mounted against the decoy fires;

since dawn, however, the camp had been quiet. Many of the enemy were visible, watchfully manning the walls.

My heart began to thump against my ribs. This engagement, insignificant though it might be, looked like being my first real battle. I felt far from how I thought an officer coming into action ought to feel. The meal I had eaten, though light, seemed to lie in my stomach like lead; my mouth was dry; and I was shaken from time to time by fits of shivering. The morning had been overcast; now, to make things worse, a steady drizzle was falling. I set my mouth and loosened my sword in its scabbard. I knew Vidimerius would be keeping a watchful eye on me; whatever happened I must try and acquit myself reasonably well. I thought, wryly, of my childhood preoccupations, but Hadrian's spirit seemed as far away as Italica itself.

To occupy my mind I ran over the main points the Duke had outlined. The attack would be three-pronged, feints at the eastern and western gates being accompanied by a thrust in strength through the low ground to the south. This Vidimerius would lead himself, at the head of the Arcadian infantry. The Galicians, lightly armed for the most part, would advance in open order ahead of the column. The Hispanians had been entrusted with the western operation, while I was to lead a hundred cavalry through the woods to the east and attempt to storm the lightest-protected of the gates. The remainder of our horse, with the Cantabrians, would be held as a reserve. Each column would have with it a party of Galician archers; if the barricades did prove troublesome they would be set alight and the attacks pressed home under cover of the smoke. In our commanding officer's view the affair was as good as over.

The column left the road a mile south of the fortress, while still concealed by the intervening spur of woodland. There the forces divided; and Vidimerius called me to him. 'I'll give you the turn of a sandglass to get in position,' he said, 'so don't waste any time. These Franks are a slow-witted pack of bastards in the main, and Alamanni are worse. If I know anything about them they'll have their whole strength posted to the south; as like as not you'll draw the entire line. If you can get in, well and good; but don't get yourselves wiped out trying. If you're held, just make a Devil of a row till we come up.' I saluted expressionlessly, wondering exactly how I was supposed to restrain a hundred bloodlusting tribesmen once they got within killing distance of an enemy. The Duke settled his helmet more firmly on to his bull-like head, buckled the heavy strap beneath his jaw. 'The Gods be with you,' he said formally. 'Though I don't expect you'll run into much trouble. If only that bloody catapult had arrived; we'd have shown 'em something then. . . .' I left him staring morosely to the south, as if he expected his favourite toy to materialise at any moment over the horizon.

While my men formed up, the Hispanians broke cover with a clatter, surging off left towards the road. The distant bawling of war horns told us they were observed. I rode to where a gap in the fringe of trees gave me a view of the fortress. It stood massive and sullen, turf battlements rising sheer over the dead ground at its foot. A surge of figures was already racing along the skyline towards the western gate.

A muttered word from one of the German decurions warned me we were ready. The column had formed in double file; a hundred burly German lancers, half a dozen slight-built Galicians at their head. Quivers were slung on their backs; from them protruded the bulky tips of fire arrows. I moved my horse forward, under cover of the trees; within moments Vidimerius and the waiting infantry were lost from view. As the first fold of ground hid us from the main column an odd silence fell.

And now began a weird experience, the memory of which remains with me strongly. The hiss of rain, the occasional jink of harness or snort of a horse, served, it seemed, to intensify the quiet. To either side rose gloomy trunks; between them the ground, denuded of undergrowth, gleamed faintly, pale as bleached bone. I rode at first as nearly as I could judge due north, keeping close to the edge of the wood; but a few hundred paces farther on the faint track I was following angled to the right, swooping from sight beneath a tangle of massive, low-hanging boughs. I followed it; and instantly the nature of the forest began to change. I realised, with an apprehensive shock, why it had appeared so inky-dark under the moon. On its edges the trees consisted for the most part of gnarled and stunted yew; here the path wound between massive cedars. Their lower limbs, thick as the body of a man, swung and loomed above my head. Under them, the gloom was intense; it was as if I swam not in air, but beneath some primal sea.

I glanced behind me. The edge of the wood was visible only as a confused brightening; and the ground was falling again, cutting off what daylight remained. I began to sweat. Once before, I had bolted from a forest; now the fear was on me again, irrational but intense. Never in my life had I seen such a forest as this.

I swallowed, and tried to concentrate on the job in hand. The ground, I knew from our reconnaissance, trended sharply downward, forming a wide gully round the south-eastern perimeter of the fortress before rising to the wooded ridge beyond. The steeper the slope, the faster we would reach the lowest point and begin to climb once more. Unconsciously I had quickened my pace; behind me the column was jostling and cantering, trying to keep up. I made myself slow to a walk, hunching my shoulders under my cloak, striving not to look up at the looming threat of the trees. It was impossible. My eyes seemed drawn, with a horrid fascination. Higher and higher they grew; and still the path plunged down. The light now was almost wholly gone; in the darkness the chalky soil gleamed evilly, stretching away in dim arcades to either hand. I panted, feeling I might choke; and the wind rose, suddenly. It couldn't reach beneath the close-set limbs; I heard it above me, a long and sourceless growl, heavy with the terror and crushing weight of the wood.

It was too much. The horse, scenting my fear, had quickened its pace again; I gave it its head, crouching low along its back, clinging with the utmost strength of my knees. A tangle of fallen debris barred the way; the creature gathered itself desperately and leaped. It cleared the obstruction; and the earth seemed to open beneath me.

That moment still haunts my dreams. Here at last was the gully I had been aiming for, but steeper and more impassable than I had ever thought. Before me was a steep-sided pit, filled at the bottom with a jumble of flinty boulders. Growing in it, reaching towards me and far above my head, was the biggest tree I have ever seen. Its trunk, yards across and black as dull pitch, thrust up beneath me, supporting plateau on plateau of dense dark foliage that rose, it seemed, to mingle with the clouds. The shock to my already overstrained nerves was enormous; what added a unique horror was my viewpoint, poised as it were on equal terms with the monster, halfway between earth and sky. I heard the horse scream, shrilly; next instant its hooves had skidded. It was sliding, scattering earth, directly down, towards those appalling branches.

How the creature ever regained the lip of the pit I shall never know. Its haunches touched the earth; it seemed to gather itself, and spring. By some miracle I stayed on its back; the next moment we were over and away, careering at breakneck speed. The column, taken by surprise, boiled to a halt as I charged back through it. I had a glimpse of a horse rearing, its rider sliding down its back; another, unable to check, plunged from sight over the steep edge of the declivity. Its shriek, and the crashing of its descent, seemed to hang behind me in the air.

All thought of the battle had been driven from me. I couldn't in any case have checked the horse; it was as much as I could do to keep my seat. I lay along its neck, gripping convulsively with my thighs. The hooves thundered on the close-packed soil of the forest floor; branches swooped decapitatingly and fell behind. I was conscious of yelling and confusion as my men swept in pursuit; then daylight gleamed ahead. Next instant I had burst from the forest edge; behind me the column, all semblance of order gone, swept into the open air.

For several seconds I was unable to take in the scene in front of me. To my right, close now, loomed the nearer wall of the fortress, its top alive with leaping figures. Beyond, glimpsed dimly through the driving rain, was the road; on it horsemen, strung out like beads on a necklace, were galloping frantically to the south. Between road and fortress, all across the tapering wedge of lower ground, a furious battle was raging. Swaying knots of men thrust and hacked at each other; horses screamed, the din of battle horns seemed to tear the air. I saw the standard of Vidimerius bobbing and waving over the turmoil; then I was carried, still at headlong pace, into the middle of the fight.

Seen in retrospect, the sequence of events is plain enough. For once the Duke, that cautious and crafty man, had fallen into the fatal error of underestimating his enemy; and in so doing had ridden straight into a trap. Now the true purpose of those night diversions was made clear. The Alamanni keeping up a running fire from the fortress were a holding garrison only; under cover of darkness the raiders had committed almost their entire strength to an ambush in the broken ground below the walls. I learned later the scouts had in fact reported enemy concentrations some two hundred paces in front of the ramparts; Vidimerius had discounted the intelligence, relying on the pace and strength of his column to break through what he considered to be detachments of skirmishers. To make the surprise complete, the Galicians had been allowed to pass unmolested; then warriors rose from the ground at the feet of the Duke and his men, cutting them off both from the infantry screen and our reserve of cavalry. The column was broken instantly in four or five places, the heaviest fighting developing round Vidimerius and his mounted bodyguard. The Germans, temporarily outnumbered, fought like demons, but to no avail; in that tight press there was no room even for a battleaxe to swing. Saddles began to empty and when I made my untimely appearance the whole outcome of the engagement hung in balance.

To an unmounted enemy there is fortunately little difference of aspect between a death-defying charge and a panic-stricken rout. The speed and weight of a hundred horsemen, striking the thickest part of the battle, blasted a path to Vidimerius. Men tumbled in all directions. I had a glimpse of a yelling face; I was unaware of the stroke I aimed, but the face split in two, becoming not human. The air seemed full of noise and the stink of blood; the horse stumbled, leaped and plunged on again. Something rang on my shield; I parried automatically, swung my sword. The blade wedged in bone, I wrenched it free, a flung spear grazed my thigh; then I was clear, and the Alamanni, confused by the irruption of horsemen from the least expected direction, were scattering across the grass. I hauled the horse round by brute force. No time for thought; I saw a man running, hewed down at him, saw him stumble, swung in my tracks again. A score of separate combats were raging round about, but the general trend of the fighting was already clear. The biggest group of Arcadians, heartened by the enemy's indecision, had burst the ring containing them; a desperate retreat had begun, across the grass to the ramparts of the fortress. None of the invaders reached it. They were caught in their own trap, for between them and safety now stood the cohort of Galicians. The little hillmen, unused to close-quarters work, hewed and thrust in sheer desperation; Franks and Alamanni, caught between two fires, wavered uncertainly. The twin impacts of the Hispanians and the

Arcadian reserve turned withdrawal to a rout. The Cantabrians never engaged; the barbarians, outnumbered in their turn, fought and were felled where they stood. The fortress garrison, preferring, it seemed, death to dishonour, plunged into the mêlée from the rear, with little or no effect; within minutes it was over and there rose from the grass that dreadful aftermath of battle, the thin susurration of wounded and dying men.

I rode slowly to where Vidimerius still sat his horse, surrounded by the remnant of his bodyguard. He was a fearsome sight. A spearpoint, glancing between shield edge and helmet, had laid his scalp open to the bone. The flap of skin hung forward above one eye; he was shaking his head, like a horse in a cloud of flies, sending gobs of blood flying in all directions. When he recognised me he gave a frantic roar. 'You,' he bellowed. 'Where in all the Hells did you come from?' Tonantius, dismounted at his side, reached to steady him, but he shoved him away. 'It takes more than a scratch from an Alaman to damage a Burgundian,' he said, swearing. 'Get a message to Merobaudes. Tell him . . . tell him the column . . .' He faltered suddenly, swaying in the saddle; then lost his balance and slid in an undignified manner to the ground. The bodyguard watched down impassively; Tonantius leaned over him, working at the strap that held his helmet closed.

I looked down. I still gripped my sword, but now the blade was red to the hilt. Abruptly, I started to shake. The field round me flickered; a wave of nausea rose in my throat. I dismounted, staggered for the nearest bushes. I was going to vomit; and I wanted to do it in private.

The tent was illuminated brightly by a collection of lanterns and tapers. Across one end of it a couch had been placed; next to it an improvised table was loaded with food. Vidimerius, paler than usual and with a mass of bandages covering his head and one eye, lay reclining on his elbow. His appetite at least seemed unimpaired. Between us stood a plate of oysters; he lifted the biggest of the shells, tilted, and sucked its contents down with every appearance of relish. I followed suit, flinching. I loathe all shellfish; always have, and always will. The Duke's one serviceable eye watched balefully while I swallowed and gulped. My stomach revolted; somehow I kept the thing down.

Vidimerius took a swig of wine, and belched roundly. 'I don't mind admitting,' he said, 'I thought for a few minutes I'd be supping at Ziu's table tonight; and I didn't find the prospect entertaining, I can tell you. Swallow it, man, don't play with it,' he interjected irritably. 'It won't bite you. . . .'

I sighed, wiping my mouth. Camp had been pitched within the ramparts of the fort, guarded now by Galician and Cantabrian sentries. From where I lay I could see through the open flap of the tent, down past the massive shoulders of the earthworks, blue now with the coming night. The sky had cleared, towards evening; the first stars twinkled in the deepening turquoise. Beyond the ramparts, in the lower ground, a wavering glow showed where a great fire still burned; it had roared all day, consuming the bodies of the slain. Nearer at hand, voices and laughter echoed; the Arcadians, drunk to a man, were noisily celebrating their success. The booty had been considerable. A dozen waggons had been taken, heaped with spoils; grain and money, weapons, armour, bolts of cloth. As Vidimerius had prophesied, the raiders had been fat; he had presided in person over the distribution of the treasure, a measurable proportion of which was now stacked within his own tent. To round off the triumph our baggage train, admittedly minus the onager, had finally put in an appearance at base camp. A hastily detached party of cavalry had returned escorting a waggon loaded in part with brine barrels; hence the oysters. The Duke could have asked for nothing more.

He skimmed a shell through the tent flap and belched again. 'Just what happened to you, anyway?' he said abruptly. 'Why did you turn? Did you hear us engage?'

I set my mouth. This was the moment I'd been dreading. I considered once more, swilling the wine round in my cup; but an answer had to be made. I could have lied; but he'd hear the truth, or a suspicion of it, from his own people fast enough. I embarked, hesitantly, on an account of what had transpired. I expected a bellow of rage; instead he heard me out. Then he shook his head. 'It's clear enough to me,' he said. 'You met a forest demon; I've known such things before.' He shuddered briskly, and made a sign with his fingers.

It was about the last thing I expected from such a professional unbeliever as Vidimerius. He caught my look, interpreted it correctly, and laughed. 'I know I haven't got much time for Gods,' he said. 'The more chanting and branch-waving that goes on, the less effective they seem to be. But demons are a different thing again. Show me the man who's never met a demon and I'll show you a bloody fool.' He took a gulp of wine. 'There was a demon in Burgundia when I was a child,' he said. 'One of my father's men spent the night in the forest, to prove his manhood. They found him later, wandering with great marks on his body. All he could tell them was, it was strong, and cold, and headless. He died just after.' He shivered again, brooding in his turn. 'Well,' he said, 'it's beside the point anyway. I'll admit I had my doubts about you when you first turned up, but this morning changed all that as far as I'm concerned. There isn't the slightest question that you saved my life, and you'll find a German isn't ungrateful.' He rose, awkwardly. In a corner of the tent reposed a heavy metal-bound chest; he unlocked it, and opened the lid. 'I saw you took nothing in the share-out this afternoon,' he said. 'You'll take this now, from me. . . .'

I know my jaw dropped. He held out, silently, a thick torque of soft, pure gold; the sort of gift he'd told me once a German chieftain might bestow on a favoured friend or follower. I had enough sense left not to argue; I took the thing carefully, slipped it on my arm. Vidimerius subsided, puffing, on to the couch. 'Look after it,' he said. 'If at a need you ever come to Burgundia, show it at any chieftain's hall and use my name: Vidimer, son of Gundobad, son of Gundieve. You'll be received as the brother of a King.'

I lowered my eyes, too confused for thought.

'For the present,' the Duke went on, 'I'm making you my second-in-command, answerable only to me. You're a good officer as far as administration's concerned; if you can fight as well, so much the better. You'll find the ring'll put you on a different footing with my people. I expect you'll be unpopular with the Latin-speaking staff as a result, but you'll have to put up with that. I think you can take care of yourself, anyway, if you have to. Now pour yourself some wine; and for Ziu's sake take that addled look off your face. I didn't ask you here to sit and stare like a stranded fish. . . .'

I made the rounds of the sentries that night, my mind in an odd turmoil. Now, with terror and confusion behind me, I at last had time for thought. The moon, high and aloof, cast my velvet shadow at my feet; below me the remnants of that terrible fire still glowed and pulsed. As I watched, a log collapsed, sending up a quick cloud of sparks. They swirled into the dark, transient as the souls of men. A bird called, in the wood; the nearest trees hung still, black fringes to the glittering sky. I was oppressed by a vague, vast sadness. Already it seemed the battle, in which so many folk had died, was a thing of the remote past, smothered, as I had been smothered in the forest, by the amassed weight of centuries. One day, inevitably, the grass would stir above my own grave; and the sun would roll, the moon float in the sky, as if I too had never been.

I pulled my cloak round me, drew myself erect. My fingers moved up to touch the rough circlet on my arm. Barbaric the thing might be, but the man who had given it was a man I had grown, however grudgingly, to respect. What had happened in the wood I didn't pretend to understand. Had events turned out differently I might have been proved a coward or a fool, but the Gods had been kind. It came to me that for the first time, perhaps, I had achieved something in which there was no shame.

Chapter nine

We campaigned in Gaul and Belgica for the rest of the season. Autumn saw us close to Germania; in early October I had my first sight of the Rhine. We rode north-west, following the course of the great river. I had heard a good deal about the fortifications that guarded it; none the less I was still amazed at their scale. Mile on mile they stretched, fronting the endless pine forests of the north; towers and walls, ramparts and ditches, frowning gatehouses and barbicans. We rested finally for two nights at Colonia Trajana, guests of the Roman garrison commander. Vidimerius, Tonantius and I, with a few of the more favoured and presentable of the Arcadians, attended a banquet he gave; the following morning the Praefectus accompanied us personally on a tour of the town's defences, the most massive I had seen. I remember him standing brooding down at the river flowing broad and silent below the walls. It was a still, warm day; the far bank stood out clear and sharp, the ranks of trees motionless in the sun haze. It seemed impossible that that silent, beautiful land could house nation after nation of implacable enemies; but our host shook his head. 'Make no mistake, gentlemen,' he said quietly. 'Here, where we stand, are the true walls of Rome. . . .'

We took our leave later that day; Vidimerius was afraid that too protracted a glimpse of their homeland might stir queer thoughts in the minds of the Arcadians. But the German foederati seemed in the best of humours. They clattered south in fine style, pennants flying, freshly greased hair glinting in the sunlight. Not a man there but was better dressed than when we had set out from Burdigala. Gold gleamed on armbands and necklaces; richly decorated capes flew in the breeze; some of the cavalry had even possessed themselves of sets of harness replete with silver and precious stones. It all made an impressive if sharply un-Roman display. The Arcadians themselves were well aware of the effect they created. In village after village peasants crowded the doorways of their hovels, knuckles to their mouths; the towns along the route, or such of them as retained a little wealth, wined and dined us royally. At Parisi we rejoined our previous road; three days' march to the south we came on the onager, weathered and gaunt, still stuck forlornly in a bog. Vidimerius, in a last lingering fit of pique, set fire to it.

By mid-November we had crossed the border of Aquitania; a few days later we were reinstalled in our old quarters. Ruricius, stouter and more debauched-looking than before, met us with a string of petty grievances. Vidimerius for once brushed aside such matters as unpaid tribute. The booty collected in the course of the campaign had been sent down ahead of us at intervals; grain stocks were secure for at least a year. The Germans settled to a well-earned rest and the nightly embroidering of taller and taller tales.

As Vidimerius had prophesied, the Arcadians' attitude to me had changed considerably. Their old air of sullen indifference was gone; when I had to deal with them now I was met with broad grins and cheerfully insubordinate bantering. I wore the gold torque prominently, and was reminded again how different are the workings of the German mind from ours. Romans would have viewed such a mark of favour with suspicion and distrust; I had no doubt at least what Ruricius' comments would be.

The long break with routine had invigorated me; I returned to my studies with enthusiasm. I renewed my friendship with Patermuthis; he was interested in all aspects of the Belgican campaign and sat with me night after night, drawing out every last detail. At his suggestion I began work on a brief account of my experiences, to which I intended to add my own observations on the use of irregulars and foederati in the general defence of the West. It was a project doomed never to be completed. In the summer of my fourth year in Burdigala, Fate struck at me again; a malignant Fate, that had followed me all the miles from Rome.

The trouble began with an official communication from Mediolanum. We were to receive a visit from an officer of the

Corps of Tribunes and Notaries; Vidimerius was instructed to make all the unit's records available for inspection and to give him every assistance in drawing up a report on our current strength and readiness for war.

The import of the directive was largely lost on my commanding officer, who knew nothing of the Corps of Notaries and cared less. I took it on myself to explain something of its functions and reputation. Its record as a government department was long and unenviable. Under Valentinian and his brother its powers to pry into every aspect of financial and military administration had given many men of dubious background the opportunity to enrich themselves; Theodosius had sharply curbed the influence of the Notaries, but they were still a force to be reckoned with. Officers of the Corps, I explained patiently, were answerable to nobody but the Emperor himself; on occasions they had been known to bypass even Praesides. If the records of the Arcadians were found wanting—as they certainly would be—the trouble could be endless. At best it would mean a new posting for the unit, to some more unpleasant theatre of war, at worst withdrawal of the foederate status itself. Vidimerius and his men could be declared outlaws, hounded from the Empire they had worked so well to defend.

The reaction of the Duke, when the significance of the letter was finally brought home to him, was a monumental outburst of swearing. As soon as he calmed down he sent for me again. Would I be so good as to set the accounts of the Praetorium in order? With my book-keeping skill, he assured me, the job would be easily done in a day or so, certainly before this spying three-legged short-arsed ape of a Roman—here he digressed into another stream of blasphemy—could get within reach of Aquitania.

The task was an impossibility. Bringing accounts up to date is one thing; creating them from a void is quite another. The amount of bookwork required, even for a unit the size of the Arcadians, is staggering; everything, from bridles and new spear-tips to the last pair of bootstraps, must be accounted for before the paternal government is satisfied that all is well. I did what I could, but by the time the unwelcome visitor was due to arrive the ledgers still presented little more than a patchwork of deceit. Any clerk worth his keep would see through them in a morning. I reported as much to Vidimerius, who shrugged. 'Oh well,' he said hollowly, 'if you've done your best, there's an end to it. Nobody could have done more; and I certainly couldn't have done as well. We'll just have to sit it out, put a good face on things and hope.'

The Imperial accountant arrived the following morning, in considerable state. He wore the full-dress uniform—breastplate, crested helmet, scarlet cloak—of a Tribune of the long-disbanded Praetorian Guard, and was attended by a whole comitatus of his own. I had been sent to the town gates to welcome him on behalf of the Arcadians, and conduct him to the Praetorium. At my first sight of him my heart sank, seeming to lie like a chill leaden lump against my ribs. I introduced myself, curtly; his eyes met mine in a narrow stare but he made no comment. I saw to the stabling of the horses, walking like a man in a dream. The sun shone brightly in the streets, but it held no warmth for me. It had been like a visitation from a ghost; in that unlooked-for face I had seen my own doom.

The Tribune wasted no time getting down to work. A room in the Practorium had been made available to his clerks; the whole party settled to their task the following day. Halfway through the morning I was sent for. Why, I was asked, were there no counterfoils for deliveries of wine and oil? I replied that to the best of my knowledge it had never been our practice to issue receipts. An hour later another query arose. Here were clearly listed an entire season's levies in grain, as allotted by the office of the Praeses; yet returns for the amounts actually received seemed radically incomplete. Did I have an explanation? I answered stiffly that certain of the day-to-day affairs of the Arcadians had been placed outside my jurisdiction. The Imperial Notary smiled at that, raising his eyebrows in mild surprise. 'Is that the case?' he said pleasantly. 'I was given to understand that after your arrival here you spent some weeks attending to nothing but the accounts. I must have been misinformed. . . .'

And so it went on, day after day. I could complain of no discourtesy; all my replies were carefully noted down, the inquisitor nodding and smiling and thanking me for my co-operation. The net was round me; each new discrepancy merely drew the meshes a little tighter. A dozen times I was on the point of going to Vidimerius, but it would have been useless. Those accounts, in so far as they existed, were largely penned in my own hand, and that the Tribune knew as well as I; I was damned, by my own excess of zeal.

The investigation dragged itself out for a week or more. At the end of that time the scribes left for Mediolanum, and Vidimerius at least breathed easier; but the reprieve was more apparent than real. The Notary remained, stalking about Burdigala in his impeccable dress; there were, he explained politely, other aspects of life in the town with which it was his duty to acquaint himself before compiling his final report. First he asked Vidimerius if he would be good enough to parade the entire strength of the Arcadians. The inference was so broad even the Duke couldn't miss it; but there was no way out of the impasse, and the muster was duly held. Our visitor strolled slowly from rank to rank, rather like a general inspecting a victorious army, while Vidimerius fumed and puffed behind. We were complimented on the smartness of our turn-out; no word was uttered on the subject of a hundred or more nonexistent men. Next the Tribune requested permission to observe the Arcadians in training. I stood beside him as he watched, with a faintly amused smile, the efforts of the German axemen and lancers. I prayed for a stray shaft to come his way, but no such miracle, of course, occurred. Finally he noted our patrol frequencies, with strengths and routings, and asked to be allowed to ride with the next detail.

As it happened there had been rumours of minor trouble to the south. Little damage had been done, but the local fear of Bacaudae had revived. Vidimerius made this an excuse for a show of strength. A hundred horse and two centuries of foot were despatched from Burdigala. I was to lead the expedition, with Tonantius as second-in-command. The night before we were due to leave he reported sick. There was nothing for it but to take Ruricius. We made a routine sweep, taking several days. We found no disorder; and I turned north again towards our base.

The trouble I had been waiting for finally came on our last day out from Burdigala. We had camped in a pleasant spot a half mile from the road, on the banks of a shallow brook shaded with clumps of hawthorn and willow; the tents had been pitched, the cavalry lines set up and the men were settling to their evening meal. A large, open-sided structure served as the Praetorium; within it, sheltered from the still-warm rays of the sun, trestle tables had been erected. As officers, Ruricius and I sat with our guest; the German decurions, who usually ate with us on the march, were also present. Among them was Gundobadius, an unwieldy and much scarred trooper who had recently been promoted following exceptional service in the Belgian expedition. Most of us had laid aside our swordbelts; only the Notary, who I realised never went anywhere without a weapon at his side, chose to sit down armed. The food was simple but excellently prepared; Vidimerius, bearing in mind the well-being of our guest, had allowed a Greek chef and his staff to accompany us from Burdigala. The table was emptied and the wine passed round before anyone spoke; then the Tribune cleared his throat.

'Well,' he said to the company at large, 'I'm sure you'll be pleased to hear I shall be sending in a favourable report. I've found things here to be in excellent shape; with one or two regrettable exceptions, of course. But those can be readily dealt with, and shouldn't affect the majority of you at all.' For the first time since we had met, his eyes dwelt on me steadily; his voice was still pleasant, but I saw the depths of malice there. Beside me Ruricius stiffened expectantly, glancing from the visitor's face to mine; the Germans, who for the most part didn't have any Latin worth mentioning, shrugged and got on with their drinking.

The Tribune drained his cup and set it to one side before he went on. 'Well, Sergius,' he said then, 'how did you find Rome?'

A vein began to pulse in my temple. I smiled, and answered as lightly as I could. 'Not bad,' I said. 'Too hot in summer, of course. Too cold in winter.'

'Ah yes,' he said. 'Good old Rome; always too hot or too cold. I did hear, though, your last year there was even warmer than usual?'

I closed my fingers on the wine-cup and stared down at the table. In a bush a few yards away a solitary bird was singing; I counted the notes individually as they fell.

'Very much too warm,' said my tormentor. He turned to Ruricius. 'Did he ever tell you,' he asked, 'how he came by that mark on his hand?'

The goblet I was holding smashed down on the table. The heads of the Germans jerked round in unison at the noise.

'Publius,' I said, 'I never looked for a quarrel with you. Don't let one begin now.'

He stared at me calculatingly with his dark, long eyes. 'Cornicularis,' he said, 'you will note which of us it was who first described our conversation as a quarrel.' He turned his attention back to me. 'It's a great pity, Sergius,' he said. 'You showed excellent promise as a scholar, once on a time. I'm sorry to see you come to an end like this; falsifying army accounts in a second-rate Gallic town.'

I spread my hands flat on the board. It had come, as I had known it must since I first set eyes on him. One glimpse of that olive skin, that arrogant, hooked nose, had been enough; but until he had spoken I had still tried not to believe. He had waited years enough to settle his imagined score with me; now, blind chance had thrown me in his path. I wondered, dully, why he hated me so much.

He had denounced me; now it was plain he intended to force me to some hasty act. He had chosen his moment well, for the only Latin-speaking witness present was Ruricius. He, perhaps, was already primed on the part he had to play. It occurred to me that maybe even Tonantius' sickness had been feigned. I had never tried to ingratiate myself with the staff officers; and Vidimerius had warned me his favour would make me more unpopular than ever. He had had no doubts as to my ability to protect myself, but he had been wrong. I couldn't protect myself from this.

Publius reached across to where my right hand lay on the table. He lifted it curiously and turned it, examining the scar. 'I gave him this,' he said. 'Many years ago, in Hispania. I hoped it might improve his manners, but it never seemed to.' He squeezed my little finger deliberately, compressing it into line with the rest. Pressure on the damaged joint has always caused me exquisite pain; I set my teeth, feeling sweat start out on my forehead, and suddenly the rage that had filled me was gone. In its place was icy calmness and a strange, arid joy. I saw what the end of the thing must be; saw, too, that it had been decreed by the Gods since the very start of time. With the calmness my brain worked swiftly once more. I was badly placed for fighting. Behind me, only a matter of inches from my back, was the stout end wall of the tent; to my right, at the end of the trestle, was Ruricius. Publius sat between me and the open air; and he was armed, whereas I was not. I needed room; I eased my foot up carefully till it rested against the trestle of the table.

The Tribune turned my hand palm uppermost. 'His gracelessness,' he said, 'extended itself eventually to Rome. Perhaps he should have fulfilled his needs in the Summoenium; instead he attempted, literally I understand, to climb beyond his station. This must be the mark he was given for his pains; it was an amusing little tale. What was her name, Sergius? Julia, was it not? Perhaps not; one laid so many women while in the city, one rather tends to forget.'

During my time in Burdigala I had picked up more than a smattering of German. Now I was glad of it. 'Gundobad,' I said softly to the man on my left, 'this officer will try to kill me. Him I can deal with, but you look after Ruricius. He must not intervene; neither must he be harmed. Or you will suffer, who are innocent.'

Publius patted my hand chidingly. 'Don't chatter to your savages, Sergius,' he said, 'until I've finished speaking.'

I swallowed. 'Publius,' I said appealingly, 'we were friends once, in Italica. Don't let enmity come between us now.' I gripped the table edge and heaved, at the same time straightening my leg.

He was taken wholly by surprise. The table, rearing in the air, sent him sprawling; he landed heavily, the trestle across his chest. Before he could roll clear I was across the tent to where my swordbelt hung from a peg. Somebody tossed me a shield; simultaneously Gundobad, with a tigerish leap, bore Ruricius to the ground. The plot, if plot there had been, was spoiled; his eyes rolled terrified above the dagger the German pressed against his throat.

Publius rose to face me slowly. Across one cheek ran a brightening weal where the table edge had caught him; he wiped at it with his palm, stared unbelievingly at the blood on his hand. When he spoke his voice was still quiet, but his eyes were swimming with rage. 'It seems your manners still stand in need of correction,' he said. 'But this time the lesson will be much, much sharper.' He drew, with a hiss of steel.

Momentarily it was as if time unreeled itself. The hot Hispanian sun was on my back; I felt the terror that had gripped me on that long-ago day, saw the pain and anger in my mother's eyes. Then he came on.

He was stylish, and fast; but he was an amateur, and the rage had made him careless. His blade rang on my shield; I stepped back, fetching him off balance. There was a moment when I could have run him through the body, and he knew it. He drew away panting, came in again more carefully.

Behind me I was aware of shouts and running footsteps. The whole camp was scurrying towards the tent, roused by the novel spectacle of two Roman officers at each other's throats. I retreated quickly into the open. The swords squealed together, but this time he didn't escape unscathed. I parried, feinted and swung my blade again. The blow, delivered with all my strength, caught him flat across the face. It was the first; and it was for Calgaca.

He fell back a pace, shaking his head to clear it. I circled, inviting him to attack. On three sides of us now crowded the tribesmen; on the fourth was the bank of the brook.

He came for me furiously again. I fenced him, watching his face, waiting my chance. The second blow was for Marcus. It fetched him to his knees; when he rose there was a new expression in his eyes. His cheek was puffy and discoloured, swelling as I watched.

He licked his mouth, glancing to right and left, but there was no escape; the Germans, silent to a man, hemmed us in. He attacked once more, desperately. I hit him, viciously, in the identical spot. This time I gave him no chance to recover. I beat him to the margin, into and across the brook. Water swirled round our calves; he retreated in panic, snatching glances behind him at the farther bank. He almost reached it before his foot turned on a stone. I could have killed him where he stood. I raised my blade; then, impelled by a sudden gust of fury. smashed the pommel down across his face. The tailpiece, curved and rough, stripped the lid from one eye, opened his nose in a leaf-shaped gash from bridge to nostril. The sword fell with a splash; he crouched against the bank, covering his face silently with his hands.

I stood over him while the mist cleared from behind my eyes. 'Well,' I said finally, 'do you wish to continue with the lesson?' There was no answer. He rocked in agony; and a thin stream of blood spilled between his wrists, coursed rapidly down one arm.

I turned away, stooped to retrieve his sword. It was my one

mistake; the last lesson Marcus had given me had been the first to be forgotten. A roar from the watchers warned me; I flung myself to one side, floundering. The dagger stroke he had aimed at the back of my knee missed me by inches; he loomed over me, raising the weapon again, and my reaction was faster than thought. The mouth was open in the ruined face, gasping for air; but the bite he took was too much for him to swallow. I thrust, off balance; the blade passed between his teeth; next instant the pommel was wrenched from my hand as I fell.

I rose dumbly, stood staring. The thrashing had stopped; he lay shaking, face down in the water, head turned to one side. The sword-tip protruded a hand's breadth from his neck; from him, waving in the current like a flag, spread a long and darkening stain.

I turned, slowly, the blood-shout of the Germans still ringing in my ears. The sun shone, its rays levelling towards the west; somewhere, idiotically, a bird still sang and piped. I stared up at the bank, and met Ruricius' eyes. His face was pale, but on it was an expression of gloating triumph.

I waded to the bank. The Germans made way for me, silently. I walked to the cavalry lines; nobody tried to stop me as I saddled my horse. I mounted, rode from camp without a backward glance. A mile away a second horseman ranged alongside. It was Gundobad. I reined and he handed me my sword. I touched heels to my mount again, stormed north on the high road to Burdigala. The trooper followed, pacing me in the deepening dusk.

The short night was nearly over when we came in sight of the town. Its streets were silent and deserted; the hooves of the horses rang hollowly from the fronts of buildings. I reined before the Praetorium, beat at the outer doors with my sword-hilt. The din roused Vidimerius himself; he came waddling from his sleeping cubicle, swearing and knuckling his eyes. His face changed as I ran towards him, the sword still in my hand. I reversed the weapon, panting, held it out hilt foremost. My voice when I spoke seemed like that of a stranger. 'I have killed the Tribune Aelius Hadrianus,' I said. 'I surrender my sword, and submit myself to judgement.'

A long pause, while he stared first at my face, then at the thing in my hand. 'Put that damned ironmongery away,' he said finally, 'and get a grip of yourself.' He bellowed for torches;

within minutes the place was a blaze of light. Lamps were lit in his office; he stamped ahead of me, crossed to a side table, splashed wine into a cup. 'Sit down,' he said. 'Get this into you. Then tell me what happened.'

I drank, fighting to stop the shaking in my body. The Duke faced me across the desk; Gundobadius stood at my back, aloof and silent. I described as concisely as I could the circumstances of the quarrel. When I had finished Vidimerius rubbed his skull like a baffled bear. 'Why, Tribune?' he asked me. 'Why should he pick on you?'

I answered that there had been an old feud between us, that we had known each other as children. He groaned at that, slamming his fists on the desk top. 'Why?' he shouted. 'Why didn't you come to me? It could have been prevented. . . .'

I said tonelessly, 'Perhaps it was the will of the Gods.'

He made an angry gesture. He said, 'Don't give me that crap. . . .' He rose suddenly, turning his back, and barked a question in German. The decurion answered stolidly. As far as I could tell his account tallied with mine. Vidimerius heard it through, head sunk between his shoulders, thick fingers twining behind his back. He was silent a long time after Gundobadius had finished; when he spoke again his voice was muffled. 'There is a bond between us,' he said. 'I do not choose to break it.'

He turned, walked heavily back to the desk. 'If I was a Roman,' he said, 'your head would be on its way to Mediolanum by nightfall, in a basket. Thank your stars I'm not.' He was quiet again, brooding in the lamplight, pulling at his beard; then he shook his head. 'I could deal with that little scab Ruricius,' he said. 'And my own people wouldn't talk. But if his commission was really from the Emperor this can't be kept a secret very long. There'll be an enquiry as soon as he's missed at Mediolanum. I can't protect you; they'd just have my head as well. Everything I've tried to do down here would be lost.'

I opened my mouth to speak, but he waved me irritably to silence. He riffled among the papers on his desk. 'Just after you left,' he said, 'I had a routine message come through from the office of the Praeses. They asked me if I could find an overseer for a lead mine somewhere down near Massilia. At the time there was nobody I chose to send. Now there is. I'm relieving you of your command, and seconding you for special duties.'

He shoved the papers aside and rose, stood leaning on the desk and staring down at me. His face looked lined and tired. 'I know nothing of this report you've made,' he said. 'Do you understand that? Nothing. Nor shall I till this officer's body is brought in to Burdigala. When I have to recognise the incident officially I shall send messengers recalling you. They won't reach you. Neither, if you have any sense, will you reach Massilia. If you vanish on the way it's no concern of mine. And may the Gods protect you. I'd do more if I could; as things stand, I hope I never see you again.'

He gripped my arm, briefly; and there was pain in his eyes. I turned without a word and left the room.

I rode from Burdigala two hours later. The Arcadian on duty at the gates saluted smartly as I passed through, bringing his spear shaft ringing down on the flags. I acknowledged him mechanically. The full magnitude of the disaster was still dawning on me. A few hours ago I had been a young man of consequence, a military Tribune with my whole career before me. Now I was ruined; shortly I would be a fugitive, with every man's hand against me. I bowed my head in the saddle and might have wept, but I was too empty even for tears.

At midday I changed horses at a way station. In my confused state I half expected a remount to be denied me, but there was no difficulty once I showed my pass. I rode on again, my mind in a worse turmoil than ever. My saddlebags were heavy; before leaving I had withdrawn my savings from the University. But where could I go? Once a warrant was issued for my arrest no Province of the Empire would be safe for me. I thought distractedly of Vidimerius' promise that I could find a welcome in his homeland. Perhaps I could curve north, cross the Rhine somewhere, lose myself in the unmapped country beyond. The thought was rejected almost as soon as it had formed. The idea was intolerable; I, a Roman citizen, dying a nameless exile. . . . It would be better, if it came to such a choice, to finish the thing at once. I touched my sword-hilt. It would be an honourable death; I would end like a soldier, even if I couldn't live like one.

I camped for the night beside the road, lay awake hour after hour listening for the hoof-beats of pursuers. None came. In the end I rose, kicked the fire to a blaze and flung on more fuel. I sat swathed in my blankets, staring back into the dark the way I had come. It still seemed impossible to believe that I was a wanted man. In my short life I had known both misery and pain, but always I had risen and taken my rest an honest man. Now my heart would skip a beat at the slightest footfall; I would never, in all my years, be free of fear again.

The black mood intensified. I frowned, pulling at my lip. Now, of all times, was when I needed my logical training most. I thought of Marcus, far away in Italica. What would he have advised? 'Think, boy,' he would have said. 'Don't panic, run about in circles. Use your head. . . .' The notion was calming; I set myself to analyse, as dispassionately as I could, the events that had led to my predicament.

Fact one was that I had killed a man. I had killed him fairly, while he still had a weapon in his hand. I had tried to spare him, that much could be vouched for; but it had not been given to me to do so. That last blow had been a reflexive act of self-defence; nothing on earth could have stayed it.

I shook my head. The end of the business was beside the point; what mattered was that I had been the first to draw. Words had been exchanged, heated words, and I had been goaded beyond any endurance; but the only witness to that was Ruricius. He would give his own version of the affair, and would certainly deny any complicity in a plot to discredit me. The Germans, even if they had understood what passed between us, would never be listened to by any court.

There, of course, was the rub. Personalities apart, Flavius Ruricius was a Latin-speaking citizen of the Empire, a long-service soldier with a clean record of conduct; the rest were barbarians. It had been cleverly arranged; the Cornicularis would undoubtedly swear that I had instigated the whole thing, motivated by a desire for revenge. The more I thought about the case, the worse it looked; certainly I had had a first-rate motive for the removal of the Notary. I regretted, now, the haste with which I had left Burdigala. Patermuthis might have helped; I had gone to his house but he had been away from home, visiting friends in the country. I rubbed my face wearily. It was no good; he would have reached, intellectually, the same conclusion to which Vidimerius had leaped. Had Publius been a civilian, I might have stood trial with some hope of a favourable verdict, but he was not. Although they performed no military function, the Corps of Notaries were administered as part of the Army, entitled to army pay, dress and privileges; so technically I had killed a fellow officer while on active service, the penalty for which was death. I groaned aloud, hopelessly. Pointless to blame the Gods; the cause of the disaster had once more been my own stupidity. This time, though, more was involved than my pride; my life itself stood forfeit.

I lay back, willing myself to sleep; but the shadows were lightening round me before I fell into a doze. When I woke, it was with a curious resolve. There was, I had seen, no way out of the dilemma in which I had placed myself. If I gave myself up it would be to summary execution, while fleeing the Empire altogether was equally distasteful. Whatever I did, then, my fate was sealed; and if death was imminent, Massilia was as good a place as any in which to await it. I had my orders, disgraceful though they might be; I would carry them out, try and meet my end when it came with something of the fortitude a Roman ought to show.

The decision, odd though it might seem, had the effect of lightening my spirits. I made better time, crossing Gaul once more by rapid stages to my destination. After the first few days of shock I ceased even to watch the road behind me. The end, when it came, would be peremptory and sudden; but I would die at the time appointed by the Gods, neither later nor before.

I reached Massilia with no untoward incident of any kind. That, I told myself, was hardly surprising; the wheels of Government were turning with their proverbial slowness. The town was much as I remembered it. Beyond the walls new buildings were being raised. I was told they were a Christian monastery; the monk Cassianus, with a band of Brothers, had elected to settle there to the furtherance of the glory of God. I made enquiries for my command, found I had to backtrack some ten or twelve miles. The place I sought lay on high ground overlooking the Via Domitia itself, the road I had travelled with such high hopes a few short years before.

Its aspect was dismal in the extreme. I saw a flat, wide compound, ankle-deep in mud. To one side was a straggle of squalid-looking huts; I made out the guardroom, beside it a taller building that despite its ruinous appearance must be the commandant's headquarters. Beyond was a group of wooden dormitories and what looked to be bath huts; beyond again, set against a slope of rising ground, stood a line of low stone structures topped by squat chimneys. They smoked steadily, their fumes adding to the greyish pall that hung over the place. The wind veered, bringing with it a sulphurous stench. I coughed and spat; then rode forward, grimly, to present myself at the gates.

The tunnel stank and dripped. The torch I was holding flickered; the flame, leaping, showed blackened balks of timber, rock walls that streamed with damp. To one side a line of men, naked save for scraps of cloth about their loins, tramped stolidly, supporting themselves by iron handles let into the wall of the cave. Beneath their feet a tree-trunk turned jerkily. Its surface was carved with a coarse spiral; the screw delivered water from a sump to a channel that, meandering to the open air, lost itself in the quagmire of the compound. A gust of air blew from the workings below; it carried with it a miasma so foul my stomach rebelled on the instant. I halted; and the man in front of me looked back enquiringly. I waved him forward and followed, drawn by a deadly fascination. A short, rough-hewn ladder was clamped to the side wall of rock. Beyond it was another, and another. They conducted me, by easy stages, to Hell.

It was a little over a week since my arrival; a week spent largely in rendering my living quarters fit for human habitation. The worst of the holes in the roof had been patched now, the bedding raked out and burned, the walls and floors scrubbed with vinegar. As soon as I could tolerate sleeping in the place I had turned my attention to the rest of the camp. Mines, as I had known very well before I arrived, were traditionally staffed by the sweepings of the Empire, by convicts and slaves; I had been prepared for bad conditions, but not for what I found. The whole camp reeked of ordure; neither bathhouses nor latrines were functioning; the only water on the place was brought up daily by waggon from a village some five miles along the road. Like many such establishments the mine was an Imperial property leased for working to a civilian contractor, in this case one Paeonius, a Massilian; I had made enquiries for him in the town, only to be told he was away on a business trip. I sat down, as soon as I had an office in some sort of order, to draft a letter acquainting him with the state of his property and urgently requesting a meeting. I gave it to the muleteer who brought up the weekly supplies, crossing his palm with silver to ensure its safe delivery; I also instructed him to let me know the moment Paeonius returned to the town.

So far no summons had come for me, either from Burdigala or Mediolanum. I was glad of the work I had found to do; it kept me from brooding too long about my own affairs. The camp was run by some thirty unsavoury-looking scoundrels drawn from every part of the Empire. They spent the majority of their time lounging about the huts; I paraded them, split them into work parties, set them to cleaning the accumulations of rubbish from between the buildings. In charge was an ex-centurion broken from VII Gemina, Baudio by name; a great malodorous sack of a man, bearded and with furtive little pink-rimmed eyes. The previous overseer, he informed me, had drunk himself to death; looking at the sordid surroundings I could well understand his state of mind. Drink, Baudio admitted with a sort of heavy satisfaction, had been his downfall too. He had quarters next to mine; I ordered his room cleaned out and soused as well, to his intense disgust. I gave him a clear choice: either tidy up his cubicle and himself or get out into the dormitories with the rest. I had already decided what line I would take with the people under me. I could fall no farther from grace; since my time was limited anyway, I would make the best use of what remained to me, which didn't include living like a pig in muck. There was nothing to be gained by discretion; I started making a noise my first afternoon on the camp, and kept on making it till I got my way.

Once the tidying-up operations were in progress I set out to acquaint myself with the working of the mine; it would be something else to keep my thoughts from my impending fate. I knew nothing at all of mining techniques; and what I saw at first left me little wiser. Crude lead, what Baudio called stagnum, was roasted from the ore in furnaces; but this was merely the start of the process of refinement. Repeated meltings and coolings finally produced crystals of pure metal, which were moulded into ingots and stacked aside for collection; but lead, as Baudio explained, was not what gave the workings their value to the State. The structures I had seen lining the hill were cupels, furnaces in which silver was extracted from the parent metal. Round them, when I visited them, was a scene of antlike activity. Sweating men scurried in all directions. Some were engaged in breaking apart the cooling clay with which the chambers were lined; others, working under the eyes of guards, stocked fresh chambers with their charges of fuel and ore; yet more worked massive bellows, directing blasts of air into the fires. The heat and noise were tremendous; while ever and again clouds of smoke and steam rolled back down on the workers, wholly obscuring them from sight. Mixed with the vapour came gusts of fumes that made my eyes stream, the breath catch in my throat. I left the area hurriedly, followed by the sardonic Baudio.

'What happens to them?' I asked when I could breathe freely again. 'Surely they can't live long in a stink like that?'

'Oh, I wouldn't say that,' he said. 'Some of 'em last for years. They get used to it. It gets them in the end, of course. Rots their lungs.'

'And?'

'And what?'

'When they die, man. What then?'

'The Empire,' he said, with ponderous sarcasm, 'finds us some more. Because if it didn't, production would go down. And that would never, never do. . . .'

It was Baudio now who was my guide. He preceded me down the ladders while round me the din grew louder. The throat of the pit was full of machinery, monstrous wheels that creaked and rumbled each below the next. Buckets attached to their rims lifted water stage by stage from the workings; it was discharged, Baudio told me—mouth close to my ear—into conduits driven through the sides of the hill above. I shouted, 'When do they stop?' He shook his head, cupping a hand to his ear. I shouted again, and he bellowed back an answer.

'Never. . . .'

I screwed my eyes shut in the thick dark, stared up again. Now I could just make out the occupants of the treadmills. They tramped in remorseless, hopeless rhythm, legs pumping, arms climbing the rotating wooden bars. Some were little more than boys; all were wholly naked. The water that rained back endlessly gleamed on shoulders and hips; their hair hung soaked and matted across their eyes. They couldn't stop, I realised, not till they were released; they would be whirled and battered by the momentum of the wheels. As I stared, more water jetted from one of them. For a moment, stupidly, I didn't understand; then I realised he was urinating indifferently into the shaft.

I turned away, found Baudio gripping my arm. I knocked his hand away, swearing, waved him forward again, stooping now to save crashing my head against the lowering roof of rock.

The light was dimmer than ever. I guessed more than I saw. Other figures, girls and women among them, moved in slow shuffling processions, bent under the weight of bulky baskets of ore. Water was everywhere; they sloshed through it endlessly, forward and back from the shaft. Snakes of hair draggled on their shoulders; their thighs and bellies gleamed pale as things long-drowned. Baudio shouted again in my ear. 'They don't like too many torches; reckon they use up all the air. . . .'

In the farthest depths I came on a tableau of death. An old man, grey-bearded and gaunt, lay stretched on the rock, blood on his throat and mouth. A younger man knelt over him, naked like the rest. He was mumbling and chanting; I saw his hand move, over and over, making the sign of the Cross. As if any God could look below the ground, and see what I had seen, and boast how he, and he only, conceived the world.

PART TWO

Chapter ten

Baudio faced me across the rickety desk.
   I had passed a sleepless night. Most of the time I spent pacing forward and back across the little room; I couldn't sit still, let alone lie down to rest. It seemed the stink of the mine infused my nostrils, its din still rang in my brain. In the small hours I left my quarters, walked the perimeter of the camp. It was silent now and dark, save for a sporadic glowing from the cupels. I stood a long time by the entrance to the pit. The throat of the shaft was closed for the night by a heavy iron grating. A torch burned in a niche; beneath it a solitary sentry snored at his post, head lolling sideways against the rock. As I stood I became aware by degrees of a deep rumbling, permeating the stillness of the night. It was a sound more felt than heard, that seemed to shake the very ground; the thunder of those hellish wheels, turning endlessly beneath my feet.
   I walked back to the Praetorium. My eyes had become accustomed to the dark; as I mounted the steps I saw, or thought I saw, a ghostly figure flick between two of the nearer huts. I spoke sharply, but there was no reply; the camp lay deserted and quiet.
   The night was warm, but I found I was shivering. I hunted myself out some wine, took it to my quarters. Sleep was farther away from me than ever; I sat drinking steadily, watching the light grow in the sky. I reminded myself, a score of times, that my fate was already sealed just as surely as the fate of those creatures in the mine. Any hour, now, my own summons would come; perhaps the courier was already on the road. Let those responsible answer to their consciences; the affairs of this world were no longer my concern.

It was useless. By dawn the wine was finished but my decision had been reached. I sat a while longer, watching the steely light spread and brighten over the compound; then I sent for Baudio.

I had made a rough list of my requirements; I ticked the items off as I talked, glaring morosely up at the overseer.

'The pit workers will be divided into three shifts. Have you a list of their names?'

He shrugged, stifling a yawn. 'Couldn't say. Somebody has, I expect.'

'Have it on my desk by midday. Also get me a team together from the off-duty overseers. I want every dormitory hut put in order as soon as possible, the work to start immediately. Do you understand?'

'Yes, but . . .'

'No buts, Baudio. How many of the huts are in use at present?'

'Two,' he said sulkily. 'The two nearest here. And a bit of the one beyond. Most of the rest have lost half their roofs.'

'That can be attended to. I want a man to ride to Massilia. He is to bring back carpenters and a tiler. If none are available we shall do the work ourselves. Is that clear?'

'May I ask,' he said, 'to what use the dormitories are to be put?'

'No, you may not. Which of those wretches in the workings are at present allowed the open air?'

He leered. 'None, officially. . . .'

'What do you mean?'

He raised his eyebrows patiently, as if explaining to a child. 'Some of the young ones, at night. The men need their relaxation.'

'Do they?' I said coldly. 'Let it be known round the camp that the practice will stop as from now. Future offenders will be put to death. If the men need their relaxation, as you put it, they must make do with each other. Or they may keep goats; I'm given to understand the satisfaction is superior.'

He gestured exaggeratedly. 'You can't run a camp like this. . . .'

'Do you question my right to command here, centurion?'

'No, but . . .'

'Then damn your impertinence, be silent. I have one answer to insubordination, Baudio. It is short, sharp and final.'

I dropped a hand to my belt; he stepped back a pace, looking alarmed. I glared at him; eventually he lowered his eyes. 'Also,' I went on in a more normal tone, 'you will oblige me by taking a bath. It's not too late to begin conducting yourself as an officer of the Roman Army.'

Something like triumph gleamed in his eyes. 'Beg pardon, sir,' he said. 'I can't do that.'

'For what reason?'

'No water in the bath-house, sir.'

'I was coming to that. Why is there no water?'

'None on the camp, sir, except what comes in the cart.'

'And I suppose the slaves drink the filth from the mine. No wonder they die like flies. Why is there no water on the camp?'

'Don't know, sir. Aqueduct just dried up.'

'Aqueducts,' I said, 'do not "just dry up", Baudio.' I made a note on the list. 'Have two horses ready at first light tomorrow. Have yourself ready as well.'

He saluted. 'That all, sir?' He was starting to look relieved.

'By no means. I saw children working at the face. They will be removed. If the Empire is so impoverished as to need child labour at least they can be found light work in the open air. As a first step nobody under the age of ten is to be employed below ground. If they don't know their age you must guess it. Guess accurately; I shall be guessing too on my next inspection. The slaves remaining in the mine will be decently clothed. Everyone will wear a breechcloth; in addition girls and women are to be provided with shifts, or at the least a breastband. See to that at once.'

He wagged his hands in alarm. 'Where am I supposed to get things like that? There's nothing on the camp . . .'

'Then send to Massilia. Any coarse material will serve, provided it is strong. If necessary they can stitch garments for themselves.'

'Most of them,' he said sullenly, 'prefer to be as they are. The water chafes 'em, and the sores go ulcerous. And if they're on the wheels they can't stop to piss.'

'Clothe them, Baudio,' I said. 'The other things will be attended to. I shall inspect the mine again in two days' time; if I find naked slaves the overseers will be flogged. Have you got all that?'

'Hope so, sir.'

'Well don't bloody well hope. Get a tablet or something, write it down.'

'Can't write, sir.'

My voice rattled the windows. '*Then find somebody who can!*' He jerked stiffly to attention. 'Nobody on the camp, sir.'

'You're trying my patience again, Baudio. Nobody? On the entire camp?'

'Not among the guards, sir.'

'Then where? Among the prisoners?'

'The slave Ulfilas, sir. The one you saw. He speaks Latin. Writes a bit as well. God knows where he learned.'

'Ulfilas? The Christian?'

'Yes, sir.'

'Good. Then fetch him out of there, clean him up and get him a tunic. He's to be here by suppertime tonight. And, Baudio . . .'

He was turning to leave.

'The man who's to go to Massilia. He's to bring back fruit, any fruit, and some fish. Whatever he likes, as long as it's fresh from the harbour. Also some wine. Caecuban if he can get it, failing that Alban or Falernian. Anything's better than that rotgut you drink here. Tell him to look snappy; I want him back by dark.'

I leaned back tiredly, loosening my tunic, listening as Baudio began to roar out orders. My eyes were heavy-lidded from lack of sleep, and I had not yet broken my fast; but at least a start had been made. The Gods willing, my one-man social revolution would rapidly begin to gather force. Human nature being a perverse and curious thing, I felt better than I had for weeks; were a summons to come now, calling me to answer for the killing, I was almost optimistic as to the result.

I sat down that night, for the first time since my arrival, to a decent board. I set to vigorously. The day had been busy, too busy for thought; but already the camp was beginning to show an altered aspect. Most of the rubbish had been cleared; in its place now stood stacks of tiles from the damaged roofs. One of the dormitories I had decided was beyond repair. I had ordered it demolished; the materials from it would be useful for patching up the others. I had nowhere near the accommodation I needed; but I hoped by the end of the week to be sleeping some at least of the mine workers above ground. My first target would be a shift rotation that would give everybody at least one night in three in the open air; later I would work out a system of incentives whereby increased production would be rewarded with extra recreational periods in the compound, perhaps even manumission in exceptional cases. I ate hungrily, my mind full of schemes; I had almost forgotten my instructions of the morning when a scuffling in the corridor outside, followed by a thump at the door, announced the arrival of Ulfilas. He was propelled into the room by Baudio and another guard. I nodded to them curtly, dismissing them; I had no doubt they would stay within call, for curiosity's sake if nothing else. I rested my elbows on the table and studied the man.

He was thin to the point of emaciation; the tunic they had found for him hung from his shoulders like a sack. His hair and skin were pale; he might once have been handsome, but the strongly boned face had long since set into an expression of brutish distrust. His eyes, grey-green and brilliant, moved restlessly, taking in the stark details of the room; the table set with two chairs, the desk in the corner, the one lamp hanging from a beam. I indicated the vacant place and spoke as gently as I could. 'Welcome, Ulfilas,' I said. 'Sit down with me, and eat.'

His reaction was unexpected. The eyes blazed at me, and deadened; he drew himself erect, fists clenched at his sides, lips compressed into a line. At first I thought he didn't understand, or that he suspected some trap. 'Come, man, you won't be harmed,' I said. 'Sit down and eat. You do speak Latin, don't you?'

For answer he spat deliberately at my feet.

I controlled myself with an effort. 'Don't be a bloody fool,' I said. 'I'm offering you your freedom, don't you realise that? *Sit down and eat!*'

Silence.

'*Baudio!*'

The door burst open with a suddenness that suggested the overseer had remained immediately outside. He ran in sword in hand, pulled up short and stared.

'Centurion,' I said, fuming, 'if that man does not eat, execute him. . . .'

A steaming bowl was placed under the slave's nose; Baudio raised his sword; and the struggle on Ulfilas' face was extraordinary to see. The silence deepened; then suddenly his determination broke. He sat, grabbing for bread and cramming it into his mouth, feeding like a famished animal; but for all his haste his pale eyes never once left my face.

Baudio presented himself, grumbling, at first light the following morning. He had saddled two horses; he stood by sullenly while I issued my instructions for the day. The first task I set my new assistant was to make copies of all standing orders for distribution through the camp. It seemed certain some of the men could read, whether they admitted it or not; those who couldn't, as I pointed out to Baudio, had better make shift to learn if they valued their skins. I saw a start made once more on the repairs to the huts; we left then, following the course of the channel that had once fed water to the place.

I was far from displeased with my previous night's work. Ulfilas, as his name suggested, was a Scythian; I had drawn from him, by degrees, the strange and tragic story of his life. Many years before, when the Gothic nation of the Thervingi had first been ferried to Thrace, they had found that far from entering a land of milk and honey they had been granted a desert to live in. No adequate arrangements had been made for their reception, and the unhappy immigrants had begun to starve by the thousand. Then had come a scandal that had appalled many Romans throughout the Empire. Lupicinus, Count of Thrace, had battened on the creatures' misery, selling them as food the carcasses of dogs. The price he demanded was horrifying; for each dog delivered, a Goth had been given into bondage. The slaves thus purchased were shipped to many parts of the Empire; one of them had been Ulfilas' mother. He himself had been born in the household of a rich Pannonian, where he had been given the rudiments of an education; later the family had fallen on hard times and both child and mother had been resold. Ulfilas had finished up where I found him, in the mines; his mother he believed to be dead. Despite the years of deprivation he had clung stubbornly to the faith instilled into him as a child; now, finally, it had been his salvation. His tale as he told it, simply and harshly, affected me strongly; when it was finished I shook my head. Lupicinus had been arraigned finally by the Senate, and dismissed from office, but the irreparable damage had been done. Nothing could ever compensate creatures like this for the shame and agony they had suffered at the hand of Rome.

'Ulfilas,' I said, 'I can see now how much cause you have to hate us. Yet you must bear with me. I cannot put the clock back, undo what is in the past; I cannot restore your mother to you, though if you wish you are free to return to her people. What I can do, and what I have pledged myself to do, is help your folk in the mine, whatever their beliefs. For that I need your help. Will you give it me, or no?'

He was silent a long time, staring at the floor. Then he raised a face that was an expressionless mask. 'I do not help you,' he said in his staccato Latin. 'Your God is not my God. I do not help Roma, who is the fourth beast to come up from the sea with iron teeth and nails of brass. I help my people, who one day will inherit the earth.'

'So be it,' I said, and called an indignant Baudio to show him to his room.

Baudio rode ahead of me now, slumped unhandily across his horse. The sun climbed, beating down strongly; after three hours or more I called a halt in the shade of an outcrop of rock. We shared the food and wine he had brought from camp, rode on again silently. The channel still stretched aggravatingly ahead of us, its bed dry and bare.

We came on the cause of the trouble shortly after midday. Close to the smallish lake that was its source, the channel had crossed a narrow, steep-sided valley on a string of stubby arches. Some idiot in one or other of the armies that had fought across the country had ordered them destroyed. The work had been performed with thoroughness. Nothing remained but broken stumps of stone; they thrust up forlornly, adorned already with a coating of rich green moss. The lining slabs of the aqueduct they had carried lay tumbled in confusion; from the northern rim of the ravine a clear stream fell mockingly, sparkling in the sunlight. A pool had been scooped out among the rocks; from it a rivulet meandered aimlessly, lost itself in a swampy morass a quarter mile away.

Baudio climbed from his horse, wiped his face and plumped down on the nearest boulder. 'As I said,' he remarked bitterly, 'it just dried up.' He nodded at the line of ruined piers. 'Who's going to put that lot back up again? And more to the point, who's going to pay for it?'

I glared up frustratedly, hands on my hips. I had seen plenty of higher arches, but these were high enough. At its centre point I estimated the channel had been some thirty feet above ground, and the total length of the spans couldn't be less than two hundred. I rubbed my face. It would take a skilled gang a season or more to repair the damage; and as Baudio had said, even if we could find the labour who would finance the job? Certainly not our absent proprietor, if I was any judge. Meanwhile our life-giving water soaked uselessly into the earth. It was a disappointing end to a hot, dusty ride.

I frowned, thinking back to something my father had once said. 'If you want to conduct water between two points,' he told me in the course of one of those interminable lectures, 'all that's really necessary is a pipe. As long as the outlet is lower than the source, the water will flow regardless of the gradients in between.' It had seemed impossible, and I had said as much. Water flowed over arches; it was an established fact. But he had had an answer for that as well. 'Since time immemorial,' he said in his clipped, sardonic voice, 'the dignity of the Empire has been sustained by arches. Therefore they continue to be built. It would be a tragic thing if the populace ever got the idea that the Roman monumental mason was losing his grip. . . .'

I walked back thoughtfully to the southern lip of the valley. There looked to be no fall between the ends of the channel. But there must be; common sense told me that. So if I built a cistern, here, dug deeply into the rock . . . introduced my pipeline, which would lead from a similar pit to the north . . . I stood frowning, lost in thought. The thing would work. It had to.

Baudio called up grumpily from his rock. 'Seen enough, sir?'

I nodded and walked down to the horses. On the ride back to camp I was abstracted, turning the whole scheme over in my mind. I could find no flaw in it; if my father's reasoning was right, I had my solution. By the time we reached the mine, the resolve was fully formed.

I would build a pipeline.

Materials were obviously the first problem. I sketched out on a wax pad a cross-section of what I thought I would need. Each pipe would end in a flange, which would fit neatly into the next; the joints could be packed easily enough, and rendered watertight. I frowned over the drawing, and squeezed my lip.

One thing was plain: if I wanted that pipeline, which I did, I was going to have to pay for it myself.

I rode to Massilia the following day, spent the afternoon going round the tileworks and potteries. Everywhere I was met with blank stares or knowing shakes of the head. Eventually I found a place willing, for a price, to supply what I wanted. The price seemed exorbitant and I said so, but there was no choice: I made a down payment on the spot, arranged to come back and examine the pipes before they were fired. A month later the first batch was completed; I rode back to the gully at the head of the motley labour force I had managed to recruit, eager to begin the work at once.

I had no experience of engineering on this scale, and no real idea how to set about the job. As a first stage I decided to cut a channel on a line between the broken ends of the aqueduct, wide and deep enough to lay my pipes. The work progressed slowly; waggon after waggon rolled to the valley edge, shed its load of piping and trundled away. By the time the last consignment was delivered the line had crept to the valley floor. I left the work to supervise the construction of the cisterns. To line them I had stone hauled from the valley bottom. Later I would roof them with slabs of stone or timber; they would serve as useful filters, trapping any sediment from the lake.

Meanwhile the pipeline was nearing completion. Its appearance did little to improve my confidence. The straight trench I had intended had proved impracticable due to the rocky nature of the ground; the thing straggled across the valley floor like a broken-backed snake. Over the last few yards the pipe climbed with increasing steepness. I caught the grins on the faces of the workers as they went stolidly on with their task. If this mad young Roman was prepared to pay good silver for their services they were willing to accommodate him, but the day water flowed uphill, I gathered, the sun could also be expected to reverse its motion in the sky.

Baudio presented himself bright and early on the morning the channels were finally due to be linked, to ensure himself a good vantage point at the fiasco that would doubtless follow. The aqueduct had been dammed where it left the lake; I stationed a party there with orders to remove the plug as and when I was ready. To communicate with them I set up a chain of signallers; when all was finally prepared I waved to the first, saw him pass the message on. At the lake they would be digging out the last of the clay that held the water back; I waited anxiously till a signal told me the flow had reached the northern cistern, which was filling. Tension grew as the level climbed towards the lip of the pipe. It reached it finally, and poured down.

At my end the supply announced its imminence by a curious hissing and roaring. Air sighed from the pipe; the workmen craned their necks; but it took a Baudio to leap into the sump, apply his eye to the orifice in an attempt to divine the source of the noise. The resulting jet bowled him completely off his feet; he was hauled from the pit half-drowned as the water rose rapidly to the sill. Within minutes it was gurgling off to the south, and I signalled anxiously for the channel to be blocked again. Until it had cleansed itself, the aqueduct must be diverted at the mine; I had no intention of letting the dirt and filth from ten miles of channel swirl into my baths.

I still remember the triumph I felt the day the water once more flowed into the camp. I had won a battle against odds; now everything, the sweat and worry, the money I had spent, seemed worth while. I saw to it immediately that stocks of fuel were laid in, the baths made available to mine shifts coming off duty. At first none of the slaves made use of the new facility; one and all suspected some new devilry on the part of the State. I promptly made bathing compulsory for the entire camp; later, as the habit spread, more and more repaired voluntarily to the bathhouses to relax in the rejuvenating warmth.

Summer was over by the time the supply was restored; the months had come and gone without my being aware of their passing. Still there had been no word from Mediolanum. I puzzled over the inaction; it seemed impossible that my offence could have been forgotten or overlooked. Eventually I put the business out of my mind. There was still a great deal to be done about the camp.

The improvements in conditions had not been without their effect. The slaves looked healthier to a man; so much so that Baudio admitted the overseers were beginning to fear a rebellion amongst the once-spiritless creatures under their command. It was a possibility that hadn't escaped me; I had new padlocks fitted to the hut doors, reinforced what windows they possessed with additional iron bars. I also reorganised the duty rotas, doubling the strength of all night guards. That in its turn left me with fewer man-hours at my disposal, which meant a further drain on my energy and time; there were days when I got to my rest dog-tired.

Every month a waggon came through from Mediolanum, in charge of a hard-faced treasury official and with an escort of Hispanian cavalry. The ingots of lead and silver were stowed aboard it after their number, weight and purity had been checked and recorded. I studied the returns anxiously. The first two showed a drop on the output of previous years; after that production began slowly but surely to improve. I won't pretend the results were due wholly or even largely to the better conditions. Under previous commandants it seemed obvious most of the silver refined on the camp had never reached the assay officer at all; I insisted on being present at the opening of every cupel and the subsequent remelting and casting of the bars, which were placed in a strong-room to which I alone held the keys. I had no intention of seeing my hard work nullified by wholesale pilfering.

Winter put a temporary end to most of my projects. It was severer than I had anticipated, with an icy, violent wind blowing day after day from the mountains to the north. The aqueduct froze at its source, and the water-cart once more made its unwelcome appearance. February brought mild, springlike weather; it also brought the first real crisis of my command. It was an unpleasant affair, the rape of a child of eight or nine by one of the overseers. The victim afterwards died from the treatment she received. I instituted an immediate enquiry. There was no doubt as to the creature's guilt; I had him summarily executed, and let it be known the next offender would be torn by bears. For some weeks after that I slept with my door bolted and a drawn sword at my side, but there were no attempts at reprisals; the camp settled down in cheerful expectation of a spectacular free show. But there were no more offences, or at least none that reached my ears; I was spared the necessity of putting the horrible edict into effect.

In early April I had a visitor. He arrived in a heavy carruca, its sides elaborately carved and gilded. My heart sank when I saw the equipage roll up to the Praetorium. I was working on the foundation of a new dormitory block; I climbed hastily from the trench, brushing my hands on my tunic, and hurried to meet it. The occupant, a short, portly man with sallow skin and grizzled, crinkly hair, descended from the carriage as I came up. On hearing his name I heaved a sigh of relief. This was Paeonius, the absent proprietor of the mine.

He was in a state of considerable indignation. He fumed and fussed behind me as I conducted him to my office, demanding to know what had been going on in his absence; and, more to the point, where was the magician who rumour had it could charm water into running uphill. I admitted my part in the proceedings; and he stared at me in distaste, seeming to notice for the first time my muddy hands and tunic. 'You don't look like a magician to me,' he said. 'You look more like a bricklayer....'

He was primarily concerned that money had been spent without his sanction. I told him, tartly enough, that it had been my own; he stared at me first in disbelief, then in blank amazement. The ledgers were produced; he pored over them for some time, frowning, before he sat back and shook his head. 'Well,' he said, 'these figures are certainly better than I was expecting. Very much better. How do you account for them? Not by building bath-houses for slaves....'

I had sent for wine; I handed him a cup, and poured one for myself. He drank, held his breath, then belched and grimaced. 'Plays the very Devil with me,' he said unhappily. 'Sometimes I swear my stomach's trying to digest itself. All brought on by worry, the whole thing. Convinced of it. Never used to be like it....'

I told him that as far as I was concerned the improved production was due solely to the better conditions on the camp. I also stressed that as long as I retained a free hand the upward trend could be expected to continue. The wine mellowed him; under the influence of a second and third cup he admitted he had business interests as far away as Burdigala. None of them, he claimed, had paid for years, or showed any sign of doing so. 'It's the state of the country,' he said. 'People just don't have money any more. Did you know that through most of Gallia coinage is literally ceasing to exist? What there is is valueless, except for gold. And people aren't exchanging that; they're hoarding it, as bullion. What the end of it's going to be I can't imagine....' He looked at me sidelong with his little dull-black eyes. 'But I needn't tell you that, need I?' he asked craftily.

I poured him some more wine, expressionlessly.

'Look at things from my point of view,' he said. He spread his hands appealingly. 'I'm not a hard man, God knows. But I have to live....' He caressed his stomach again. 'It's never worth it,' he said. 'I sometimes wish I'd been born a farmer, or a peasant. When I do make a bit, enough to make ends meet, they won't let me keep it. Who do they come to if the town hall roof falls in? Me, I tell you, every time. And look at that contraption I go round in. That was the Domina's idea, not mine. I hate it. Can't even drive it. And those horses. Ravenous great brutes. They don't eat hay, they eat money. Solid money, by the pound; I swear it....'

It struck me he might have enjoyed his life even less had he been born a slave; but for once I held my peace. By the time he rose to leave his indignation was forgotten. 'You're doing a good job here,' he said, squeezing my arm confidentially. 'A good job. Why Mediolanum couldn't send me a competent engineer before, instead of the jail sweepings I've had to put up with, I shall never know....' I accepted the compliment silently. My knowledge of engineering was minimal, but if he chose to think of me as an expert, so much the better. I saw him to the carriage; he hauled himself aboard, puffing, reached down to grip my arm again. He would, he told me, be away from Massilia for a further week, maybe longer; on his return he would be obliged if I would grace his home with my presence for dinner. There were, he said, many subjects that might be discussed to our mutual advantage. Minutes later the carruca swept grandly through the gates, dwindled in the distance in the direction of the town.

Overall I was pleased by the outcome of the interview. There was something nearly engaging about the little spectabilis, with his perennially delicate stomach and absurd blend of dignity and brashness. I dismissed the invitation from my mind; I was accordingly the more surprised ten days later to receive a letter, couched in flowery terms, bidding me forth to the mansion of Paeonius the following afternoon. Under the circumstances I obviously had no course but to comply; I gave my acceptance to the slave who brought the note and rode into Massilia next day determined to make the best of what I was convinced would be a typically dreary provincial occasion.

The house of the mine-owner stood almost in the centre of

the town. At first sight it was an imposing-looking place. A deep portico fronted it; within, ranged to either side of the heavy doors, stood a pair of racing chariots, their shafts propped against the wall. It was an affectation I hadn't seen outside Rome. On closer inspection the massive relics proved to be battered and weather-stained; had either of them been coupled to a team I wouldn't have backed it to finish a lap. I shrugged, and gave my horse into the keeping of a slave. I was conducted to the baths by the doorkeeper, a burly African; his livery of dark red and gold had the same air of somewhat overblown magnificence. I had arrived rather unfashionably on time; I compensated by spending a luxurious extra hour over my toilet. I finally emerged to find my tunic ready for me, freshly pressed and warmed. The Negro appeared once more to lead me to the triclinium.

The house was similar in plan to the house of Patermuthis in Burdigala, but there the resemblance ended. My mentor's home was graceful and airy; that of Paeonius was sombre and dark, full of a chill dampness that no effort of air and sunlight seemed likely to dispel. I padded through room after room, their walls stippled with frescoes in crimson, ochre and black. Busts, all in the same stolid provincial style, glowered half-seen from alcoves; between them curios from Egypt and the East testified to their owner's taste and former travels. I saw richly woven rugs, inlaid cabinets and chests; a statue of Anubis, the dog-God of the Nile, executed in some glistening black stone, another of a Pharaoh, seated stiffly with his golden crook and flail. His eyes watched with a species of blank menace as I passed. Ahead of me the sandals of the African whispered on the flags. I found myself watching, sidelong, the shadows that leaped and crowded after us as we passed.

The triclinium itself presented a welcome contrast. The many pendant lamps with which it was lit bathed the little chamber in a warm glow. Drapes covered the walls; to one side the three couches—everything, as I had expected, was classically correct —were grouped round a series of tables already laden with food. The other guests had assembled; I was greeted courteously by Paeonius, and bidden to my place. The spectabilis had arrayed himself for the occasion in a costly looking tunic embroidered with patterns of silver and gold. Rings gleamed on his fingers; on his head, somewhat comically, was perched a sagging wreath of green leaves. I sipped the mulsum that was passed to me while I was introduced in turn to the Bishop of Massilia, attending with some lesser dignitary, three local business friends of Paeonius, and the wife and daughter of my host.

The Domina Papianilla I decided I disliked on sight. She was a woman of less than middle age, running to thickness in the body and with a powerful, ugly face. Her eyes, wide, protuberant and icy-blue, were scarcely enhanced by thick-painted fringes of black; the mask of cosmetics, coupled with a strongly convex forehead, imparted to her a permanently feral expression. She wore not her own hair but a yellow wig, a full blonde scalp piled into a complicated mop on top of her head; it sorted curiously with what remained of her natural complexion. Her fingers, like those of her husband, were crusted with jewels; they moved restlessly, now gripping the stem of a goblet, now beating irritable, nervous tattoos on the inlaid table top in front of her. Just such noble creatures I had seen thronging the amphitheatre of Rome, yelling themselves into orgasms at the sight of blood.

The skin of the girl who sat beside her was clear and golden, unhampered by powder or paint. Her hair, simply dressed, was long and dark; it framed an oval, delicately boned face set with wide eyes that had in them something of the uncertain shyness of an animal. Above them her brows, in defiance of the dictates of fashion, were level and thick. A plain white robe, unadorned by jewellery, set off to perfection the warm brownness of her throat and arms. I drank wine again, slowly, remembering where and when I had seen her before.

It was in the market-place of Massilia, the day I came to the town looking for someone to fire me a set of pipes. Then as now her hair had been brushed till it gleamed, secured by a fillet behind her little, perfect ears. A slave attended her; I saw she pointed and nodded at what she wanted, leaving the haggling and buying to her companion. The wind blew, pressing the thin fabric of her robe against her, showing momentarily the swaying curve of back and hips; showing too an ankle as brown as her cheek. Then as now she had been aware of my gaze; then as now the blushing had started, spread from her throat to the roots of her hair.

But the gustatio was over, the first dishes cleared away; and

Papianilla was addressing me directly. 'This,' she said, in a husky, carefully modulated voice, 'must be the great magician. Tell me, Tribune, are you really as much at home, as my husband assures me, with bricks and mortar as with the bewitching of streams?'

I felt my colour start to rise at the delicately contrived sneer. 'Both gifts were given me by God, Domina,' I said sweetly. 'It is my burden to employ them as I can. . . .'

She smiled, graciously, the expression stopping short at her eyes. The exchange set the mood for the conversation. I was assailed from all sides by questions about my birth, upbringing and career; I answered evasively, careful to remain courteous, reserving the stray barb for Papianilla. I was, it was plain, a social curiosity, a very odd fish indeed to have blundered into these folks' little net. Paeonius watched the whole charade worriedly, casting sidelong glances at his spouse. At one point the Domina assailed me in Greek. I answered her as urbanely as I could. I am no linguist, and my accent has always been bad; I was comforted by the realisation that hers was worse. The talk switched to literature; I presented, somewhat unexpectedly, the phenomenon of a plumber who knew his Horace. Papianilla retreated, re-engaging on the subject of the fine wines of Italia. I countered with some nonsense of my own about Hispanian vintages; my father had always kept a good cellar and I knew the names and quality of most of our local growths.

Dish succeeded dish, the wine flowed more and more freely, and I began to enjoy myself in a rarified sort of way. This, after all, was a game I'd played before, many times, in Burdigala. The wine mellowed me; so much so that when I realised Paeonius' daughter had so far contributed nothing to the conversation I determined the dumb should be brought to speak. 'I saw you in the market-place some months ago,' I said. 'Shopping, with a slave. Naturally, I couldn't address you; will you talk to me now, under your father's roof?'

Instantly the room was silent. The Bishop coughed embarrassedly and addressed himself to his goblet; Papianilla raised her eyes briefly to the ceiling, as if calling on God to witness some enormity, while Paeonius stared at me with the oddest expression of mingled anger and pain. His daughter spread her hands flat on the table, head lowered, forehead suffused once more with crimson.

Certainly the wine had hold of me, for my good humour changed instantly to irritation. If I had assaulted the wretched girl in full view of the company they could scarcely have looked more appalled; as it was, all I had done was ask a civil question. I repeated it deliberately, the words falling into the quiet like so many stones. She raised her eyes at that, no longer able to ignore me. Her head rolled, agonised; she seemed to pant for breath. Her lips parted, but all that emerged was the gross beginning of an enormous stammer.

I shall never understand what happened next. Certainly her beauty and distress worked powerfully in me, aided by the wine; but to act as I did I think I must have drunk myself temporarily out of my senses. I sat up and leaned to grip her fingers, willing her to speak. A wait, while she fought for control; then she answered quick and low, in a voice like the rushing of wind in trees. *'I saw you too,'* she said. *'You were wearing a cloak. I was buying oranges.'*

The reaction of the company to the simple speech was even more remarkable. Paeonius' brows rose towards his hair while a look of amazed delight spread over his face; Papianilla's eyes widened even further with astonishment while the Bishop and the rest burst into a babble of congratulations that sent their victim even redder than before. For my part, it was obvious some prodigy had taken place; I lay back filled with a curious mixture of emotions. Had I not been in my cups I would probably never have addressed such an exquisite creature directly at all; now, knowing what I did, I wondered if I would ever have the courage to speak to her again. The answer, I knew instinctively, was yes; for something had passed between us in that moment of pain that defied ready analysis. I knew myself incapable of love; what I had felt then could have no name. Its very formlessness perhaps rendered it more powerful; for when the company finally went its several ways, in the still hours of the morning watch, I carried back with me something more vivid than a memory. I lay in my bed, feeling the strangeness of the wine ebb from my veins, hearing her name again, Paeonia, hearing her voice.

Chapter eleven

A few weeks later I had the pleasure of viewing Massilia from what was for me a new vantage point. I reclined in the rear seat of the carruca, lolling grandly against the cushions, while the ungainly vehicle lurched and rumbled its way to the west gate. Beside me, wrapped in a light travelling cloak and with a wide-brimmed straw hat drawn low over her eyes, sat Paeonia. A female slave attended her. A second retainer controlled the carriage, flicking and cracking his whip to clear a way through the busy streets; two others, clad in Paeonius' outrageous livery, cantered haughtily in our wake.

We clattered out in fine style along the Via Domitia; a few miles from the town the carruca left the highway for a rutted track that wound through low hills to the sea. There, in a secluded bay, awnings were set up, rugs and cushions scattered beneath. The horses were unharnessed and picketed; the slaves sat stolidly on the beach, respectfully out of earshot, while Paeonia and I lounged and talked. The sun beat down strongly, sparkling on the waves that lapped the fine white sand; a breeze rustled the stiff-leaved shrubs behind us, scarcely louder than her voice. Later, in the deepening twilight, we walked the edge of the sea. Our footprints showed dark against the sleeky glistening sand; she ran and laughed, lifting her robe to her knees, stooping to snatch up little shells the retreating waves laid bare. It was full night before we returned, jingling through the streets of Massilia to her father's door.

This extraordinary state of affairs had come about, as I judged at the time, largely by chance. For some days after the dinner party I brooded round the camp, unable to concentrate on anything. The image of the girl, shut away in that gloomy house, haunted me persistently. Time after time I put it out of my mind, but it invariably returned. I reminded myself bitterly of my own precarious state; lacking friends, money and influence, my life itself staked on the whim of some Mediolanian clerk. Besides, I had seen enough of women during my stay in Rome. Paeonia's world was closed to me, for all time; I told myself it was better that way.

In the end I buckled down to my tasks again, finding relief in the sheer physical effort of running the mine. The assay officer put in his scheduled appearance; hard on his heels, and most unexpectedly, I had another visit from Paeonius.

His manner was considerably altered. He greeted me pleasantly, asking if it would be convenient to inspect the workings. I roused out Baudio and a couple of overseers and accompanied the party myself, helping the portly spectabilis down ladder after ladder into the noisy dark. Conditions were much improved, though there was a lot I realised I would always be ashamed of. I pointed out what still needed doing; Paeonius agreed absentmindedly, throwing the odd nervous question about security measures. There was, he pointed out, a great deal of valuable equipment in the mine. I bore his comments indifferently; I had a growing conviction that the mine and its operation were not the sole purpose of his visit.

In that I was right, for later, over a glass of wine, he finally wheeled ponderously to a subject closer to his heart. And a grim enough tale he unfolded, in all truth; of rape and pillage, the sack of a villa in the lawless times of Maximus, a child dragged bloody and hysterical from beneath her mother's corpse. 'The Domina Papianilla,' he said in his fussy way, 'is, of course, the girl's stepmother. I . . . ah . . . remarried basically for Paeonia's good. I thought perhaps . . . another woman . . . but it's done very little. Very little . . .' He shook his head sorrowfully. For a twelvemonth, he told me, the little creature had neither washed, dressed nor fed herself unaided. Doctor after doctor had confessed himself baffled, and the spectabilis had become resigned to the prospect of living with an idiot for a daughter. But time had worked the cure herbs couldn't achieve; with the years Paeonia recovered by slow degrees, to grow into the lovely young woman I had seen. Save in one respect. When, finally, she spoke again, it was with the dreadful, stumbling rush that I had heard; the Devil had taken her tongue, and never let it go.

'Until that night,' said Paeonius, 'she never uttered a sentence that could be understood. . . .' He flashed me one of his quick, crafty glances. 'The Bishop,' he said, with an unexpected access of humour, 'would in no way have been averse to the proclamation of a miracle, had the laying-on of hands been performed by his good self. As things were, he was sufficiently impressed; he urged me to make every effort to draw such an evidently worthy young man closer to the bosom of the Church.'

He hadn't, even now, come to the point. I waited expressionlessly, but despite myself my heart was beginning to race.

He had noticed, he said—it could scarcely be missed—a certain . . . ah . . . mutual interest, an interest he for one would by no means be averse to fostering, solely for Paeonia's good. A shadow had been lifted from the household; could it be banished completely, I would earn at least his undying gratitude. He would even put his carriage at my disposal, were I to call at his home again; it would benefit the child to get some trips to the countryside, into the fresh air. So it came about that, partly against my will and certainly against my instincts, I was drawn into an involvement that seemed to promise nothing but misery and pain. Had I known what would in fact result I think I would have fled Massilia, and the Empire, there and then.

Meanwhile, though the future might be black the present was undeniably sweet. Whatever hours I could spare from my self-imposed duties at the mine I spent with Paeonia. Sometimes our meetings took place at her father's house. She would sit in the walled garden of the peristyle, a book on her knee or a piece of embroidery; her hands moved steadily and deftly, now pausing in the sunlight, now darting like brown butterflies, while I talked and talked. Of ancient things for the most part, stories of the Gods of Greece, tales of Rome and her Emperors and Kings; childhood fantasies that stubbornly had never left me, and now at last found a ready audience. While the God-figures watched from their niches, less disturbing now in the bright daylight, the summer air moved gently in the court. Sometimes Papianilla would pass through the corridor that bordered it, inclining her head to favour me with the fixed and icy smile that served as greeting between us; or Paeonius would take time off from his ledgers and files, sit fanning himself and grumbling at the state of the Province and Empire, the injustices that beset him on every side. But mostly we were left alone.

Sometimes I would go with Paeonia on her shopping trips to the markets or the harbour. She liked the harbour, with its endless bustle and confusion, best of all; she would turn over the glistening, still-wriggling fish as they were landed in their baskets, pointing and frowning, shrugging or shaking her head while I haggled with the boatmen. Her judgement was shrewd; in time we developed a whole range of signals, comprehensible only to us, by which she conducted her affairs. She would rub her nose or touch her hair, fold her arms to clinch a purchase, tap her foot on the quay to break off a deal; it was a game she never tired of. Other times I would take the carruca, with three or four household slaves as bodyguard, ride inland or to the sea. Those were days when time itself seemed to stand still; when I could forget, in spite of myself, the sword that still hung over me by a single hair.

Paeonia's affliction did not noticeably diminish, at least at first. Time after time the rolling of her head, the panic and sudden misery in her eyes, warned me what would follow; and I would touch her hand or lay a finger lightly to her lips, willing her to patience and quietness. She would swallow and pant, twining her fingers and lowering her head; and her eyes would clear, she would speak quickly with that same rushing intensity I had first heard, say her thought and be done. Then she would laugh, or we would laugh together; and once she ran and skipped, clapping her hands for joy at the new things her tongue could achieve.

So the summer passed, swiftly and pleasantly enough. Autumn brought fresh rumours of unrest in the north. All along the frontiers of the Empire the tribes were once again in a ferment; the field army campaigned endlessly, force-marching from point to point, suppressing this or that threat to the safety of the West. Also it was said the Scythians had taken the field. Imperial messengers rode the Via Domitia day and night, but for us in Massilia reliable news was hard to come by. Till one dawn Baudio roused me in what was for him a state of considerable excitement. He told me the regent Stilicho had passed in the night, heading west, some claimed to Britannia herself; with him had been the Standards of Vidimerius.

The news woke all my slumbering restlessness. Obviously the Magister Militum could no longer afford to let crack troops languish in garrison duty so far within the Empire; had it not been for the chance encounter that had ruined me, I might have been riding with him. I cursed my lot bitterly, quite forgetting that but for the inexplicable kindness of Fate I would most certainly have been dead as well as disgraced. I even conceived the childish notion of riding west myself on the track of the Palatini, but it would have been useless. Vidimerius had intimated plainly enough what I could expect if he ever set eyes on me again; I would merely be rushing on my fate.

By mid-morning the need for some sort of positive action had become imperative. I had spent an hour inspecting the results of the last season's work, and small enough they were in all faith. The footings for the new dormitories had been laid—mostly by me—but there the job had ground to a halt. In my first flush of enthusiasm I had sworn to carry the work out with the staff at my disposal, but one attempt to marshal overseers and off-duty slaves into a construction gang had convinced me of the impracticability of the scheme. I needed timber in quantity, and skilled men to shape it; the cost would exceed anything I could afford. I had already applied unsuccessfully to Paeonius. He was, he claimed, sinking deeper into debt with every year that passed. There was no money anywhere in Massilia for projects such as mine; they would simply have to wait for better times. If better times ever came. I made a last round of the litter of trenches, still fuming, then saddled a horse and headed for Massilia.

I didn't enter the town. Within sight of the walls I turned off the road, urged the animal along a steepening track to where, set on a slight promontory, the stone of the new monastery gleamed in the sunlight. A notion I had been harbouring for months had hardened into a resolve. I would seek help from the fountainhead of charity, the Church itself; where there was money for a project like that, there was money for my simple huts. I disliked the thought of begging, but my feelings were, after all, a secondary concern; God knew the cause was worthy enough.

The buildings of the place stood simple and four-square, facing Massilia and the sea; round them at a distance a high wall shielded the inmates and their devotions from the gaze of the herd. I presented myself at the gate, to be met by an elderly man whose coarse, plain robe and shaven pate proclaimed him a Brother of the Order presided over by Cassianus. He took my name, courteously enough, and enquired my business. I told him I was the officer in charge of the Imperial mines, and that I urgently required advice concerning the welfare of my slaves. He raised his eyebrows slightly but made no comment. His spiritual Father, he explained, was at his devotions, and couldn't for the moment be disturbed; but if I cared to wait he was sure he would be pleased to receive me. A second monk saw to the stabling of the horse; I was shown to a small, square cell, empty save for a stone table on which stood a loaf of bread and a pitcher of water. The door closed softly behind me, leaving me alone.

I walked to the one narrow window. It looked out across the grounds of the place. In the distance, masons were working on the outer wall; the ring and tap of a trowel came to me faintly, muffled by the thin greenish glass. Nearer at hand, a man hoed a vegetable patch with slow, methodical strokes, stooping from time to time to pluck and toss away a weed. From somewhere within the building itself came the deep-toned chant of voices, interspersed now and again with the ringing of a handbell. There was an air of security and peace; the hot, restless world I had left seemed suddenly very far away.

I was roused from my brooding by the opening of the door. The man who entered was tall; as tall or taller than I. Like the rest of his Order he wore a plain brown robe, caught at the waist by a girdle; on his breast, suspended by a necklace of beads, hung a massive cross. His face was thin, strongly yet delicately boned and enlivened by a pair of brilliant light grey eyes. They studied me for a moment before he spoke. 'I am Cassianus,' he said simply. 'You are welcome to my House, Tribune. I understand you have some pressing business to discuss; but before that, would you care to break bread with me? The Domitian Way is both dusty and tiring, as I know very well.'

I agreed, thanking him; for I had ridden the ten miles or so non-stop, and hadn't yet eaten that day. He inclined his head gravely and turned to precede me, sandals shuffling on the new, pale flags.

The monastery was airy, spacious and cool. I followed my host down a tiled corridor, its cream stone walls untouched by plaster or paint. We crossed a hall lit by narrow windows. Kitchens opened from it; as we passed a Brother emerged carrying a platter of bread. He stood back respectfully at sight of his superior. I glimpsed what looked to be a wine store, the lines of tall jars standing neatly in their racks. Beyond again, an antechamber led into the refectory.

The Brothers took their meals at wooden benches set down the sides of the wide, low room. The hum of conversation ceased momentarily as I entered. At one end of the place tall windows of the same greenish glass admitted a bursting of sunlight. Beneath them, raised slightly on a low stone dais, was a table set a little apart from the rest. I saw there was to be another guest. He was already seated; a well-built, silver-haired man with a fleshy, aquiline nose. His clothes and bearing marked him as a person of consequence; I guessed him to be the head of some local noble family. When Cassianus introduced him I had a shock. 'Allow me to present,' he said in his quiet, cultured voice, 'the Senator Prudentius Aurelius Clemens, a good friend of our Order. My Lord, the Tribune Sergius Paullus. You may have heard,' he said, turning to me, 'of the Senator's work. He is the foremost poet of the Christian cause, and a pillar of our Western Church.'

I swallowed. This was the man whose diatribes I had long ago wished might be painfully expunged; of all people he was about the last I had ever expected to meet, least of all in a Gallic monastery. I greeted him formally, admitting that not only had his fame reached my ears but that I had studied his writings while at school. Prudentius, who had been watching me somewhat narrowly, beamed at what he construed to be a compliment and lost no time in plunging into a discussion of his current projects. 'I am preparing,' he said, 'my most ambitious work to date; a work that will prove—logically prove, mark you, in the very tongue and accents of his deriders—the singularity of the one true God and of his Son, given to us to be the redeemer of the world. With the publication of the *Contra Symmachum* the voice of that capering bigot, that unbeliever, that mouthpiece of Satan himself, will finally be silenced. Praise be to the Lord, who gives us the strength to work his divine will. . . .'

There was more, all in the same high vein. I addressed myself to my meal while the Senator continued his speech, to which Cassianus responded from time to time with murmured compliments on the fineness of the poet's work and his untiring zeal.

The food was plain but nourishing: meat, wheaten cakes and fruit, washed down with rough, well-watered wine. I ate hungrily, wasting no time in talk; I couldn't in any case have stemmed the flow of the great man's oratory for long. The platters were cleared away before Cassianus turned to me to enquire gently the purpose of my visit. I explained as briefly as I could the conditions I had found on taking over the mine. I described the improvements I had been able to make, stressing the great amount that still needed to be done, and ended with an appeal for funds, however slight, that would allow the work to continue.

The reaction of the listeners was hardly what I had anticipated. The monk frowned thoughtfully, looking troubled; while Prudentius, who had begun to evince mounting impatience, appeared suddenly to be on the verge of apoplexy. 'Father, this is intolerable,' he burst out. 'Quite intolerable. The Church exists, under God, to the furtherance of his glory; are we to succour those who in their wilfulness and blindness deny their Lord, who in their evil counsels prevail on men to take his name in vain?'

I could only stare at him blankly. 'But,' I said finally, 'I think you must be mistaken, sir. My slaves are Christians.'

He glowered at me. 'They are heretics,' he said. 'Subverters of the divine will, to whom neither mercy nor compassion may be extended.'

'And I tell you you're wrong,' I said, getting annoyed in my turn. 'But even if it were true, there are children to consider. Am I to understand the Church withholds its charity even from them? Are they to suffer regardless, for sins they're not old enough to understand?'

That brought him to his feet, fairly stamping with temper. 'Don't presume, young man,' he said, 'to lecture me in my Christian duty; or, what is worse, to criticise the edicts of the Faith. A Faith of which, it seems to me, you stand in most alarming ignorance. Is not the foremost of your band a Scythian freedman?'

Cassianus interrupted before I could answer, holding up his hands for quiet. 'Gentlemen,' he said. 'Gentlemen, please....' He turned to me. 'I'm afraid,' he said, 'this freedman, the Goth Ulfilas, has been causing some slight dissension in Massilia. You were perhaps unaware of this?'

But the Senator's indignation would not be bottled up for long. 'Dissension!' he bellowed, bouncing up again. 'Dissension! Dissension and worse, Father, dissension and worse! I saw the wretch myself, and heard him, preaching his filth on the streets in the plain light of day. I'd have given him something for his trouble, I promise you, if I could have come up with him; I'd have thrashed him within an inch of his life. But the creature was too quick for me . . .'

Events seemed to be rapidly leaving me behind. 'But,' I said, baffled, 'Ulfilas is a Christian. I've heard him speak myself, to the slaves at the camp . . .'

'He is a heretic!' thundered the Senator, hammering the table with rage. 'A follower of the excommunicant Arius, a spreader of sedition, a rejector of the Son of God himself. I myself heard him proclaim that the rich shall be brought low, that cities—good Roman cities, Christian cities—will be put to the flame, that barbarism—all that is chaotic, and of the Night—will conquer, and reign supreme!' He raised his eyes to Heaven, represented temporarily by the refectory roof. 'O God of our fathers, hear thy faithful servant; visit the unbeliever with thy living fire!'

It sounded like Ulfilas, right enough, but by this time I was too angry to remember the original purpose of my visit. Helping those wretches at the mine had cost me both money and labour; now here was this fifth-rate panegyricist quibbling technicalities I barely understood, using them as an excuse to refuse aid to people outwardly Christian, who had presumably offended his sensibilities on some minor point of dogma. The insolence of the whole thing practically took my breath. 'That was most remiss of him,' I said when I could once more make myself heard. 'Be assured I'll take steps to correct his error. But,' I added, 'viewed dispassionately, perhaps his mistaken concern for the poor is understandable. Our Lord, after all, chose to make his appearance on earth in the guise of a carpenter; maybe it would have been better all round had he been born to the clarissimate. . . .'

'*Tribune!*'

This time Cassianus was on his feet. 'This is my board,' he said, fixing us both with his piercing stare. 'And you are under my roof. By your leave, my friends, we'll have an end of this.'

Silence fell. Prudentius glared at me, blowing through his

nostrils like an overworked horse; I glared back, equally badtempered. Those of the Brothers who had turned their heads at the sound of raised voices readdressed themselves hurriedly to finishing their meal. A further pause; then Cassianus turned to me. 'Tribune,' he said, 'I'll have a word in private with you, if you please. Forgive us, my Lord.' To one side of the dais was a low door; he stepped to it, motioning me curtly to precede him. I walked through silently. Beyond was a small room, furnished in part as an office. Cassianus closed the door behind him, leaned against it pursing his lips and shaking his head. 'Well, my son,' he said finally, 'for a seeker after charity you have some curious notions of how to achieve your aims.'

I was still bitterly angry. I held my hands out, showing the calloused palms, the cracked and broken nails. 'I've been told, Father,' I said, 'that Faith without works is dead. I've worked...'

'Peace,' he said. 'Be still. We are none of us strangers to toil.' He walked to the window of the place, stood with his arms folded across his chest, chin sunk forward as if in thought. 'It's fortunate,' he said at length, 'that I have a certain regard for sincerity, however oddly it chooses to express itself. You must learn to exercise a similar forbearance. Enclosed though we are, we are by no means unaware of the world around us. I've been hearing a great deal about your activities; I could wish that you had come to me sooner.'

I snorted. Prudentius struck me as a true author of his poems, a sententious bore; I muttered something to the effect, but the Father shook his head. 'The Senator,' he said, 'labours under a weight of Faith; and Faith, Tribune, can be a sore burden to the best of us. This you still have to learn. Now listen, and cleanse your heart of anger.

'In what he said, the noble Prudentius was perfectly correct. Your freedman is, unfortunately, a heretic; a follower of Arius, who preached the indivisibility of the One. His very name might have suggested it to you; for Ulfilas was a disciple of Bishop Arius, and carried on his ministry among the Goths. An energetic man, some say demoniacally inspired, who finally succeeded in converting the entire nation to his master's beliefs. Beliefs they still for the most part hold.'

I eyed him wearily. 'But all this happened centuries ago,' I said. 'Divisibility, indivisibility... Father, can you honestly say it matters? When just a few miles away Christians are dying for plain lack of air and sunlight?'

'They are not Christians,' he said quietly. 'Try to understand...' He turned back to me earnestly, ticking off points on his fingers. 'If God the Eternal is indivisible, then logically he created the Son not from himself but from nothingness. From the void. If Jesus, our Saviour, was not of his father's substance, then God never intervened physically on this earth. We restrict his authority to the matters of the spirit, and query the right of the Church to interdict the affairs of mammon. But if we accept, as we must accept, the higher mystery—that the Word was made flesh, and lived and died among us—then we see that here on earth God did indeed speak his will. We see that through his disciples the Word flowered and grew. We see that through an unbroken line—the line of the Bishops of Rome, each of whom inherited the mantle of the holy Peter—that Word continued to be uttered, and is uttered today. And that to flout it is disaster to the soul, for Kings as well as for commoners. This is the very touchstone of Faith; for on this the Church bases her right to rule.'

I stared at him. 'To rule, Father?'

He nodded. 'You heard me well enough, I think.'

I was silent. I understood, dimly, the nature of the heresy he had outlined, but its political implications had never struck me till now. The anger was gone; I saw, in a flash of comprehension, how the Church, rich as she already was, would continue to grow in power and authority, till her word stood above the once-divine will of the Emperors themselves. I seemed to sense, on the instant, a dim and mighty conflict; clergy and the State, Church and Emperor, contending everywhere for the very destiny of man.

He realised I had understood. 'Yes,' he said, 'the battle has yet to be fully joined; but joined it must be, one day. It was ordained, from the first; when a Roman Emperor gave the Church authority, to use or abuse as she chose. The burden of power; a burden she had to take up, or herself wither and die. But what the end will be, passes our imagining.'

He turned back to the window. 'For this reason,' he said, 'when we saw the destiny God had laid on her, our little Order drew apart. For this reason others will draw apart, in future years. Sometimes in our watchings we pray for the souls of men;

sometimes for the Church herself. That she may be guided rightly, and not fall into sin more deadly than Arius could conceive.'

He stayed still a moment longer, brooding through the glass. 'You perhaps see now,' he said, 'just how you played with fire; and why your fingers were so sharply burned. The Senator sometimes speaks from his heart rather than his head; hardly, as you will concede, a mortal sin. But whatever he may lack in understanding he makes up for in zeal; you'll find him a true mouthpiece of Holy Mother Church.' He crossed to me, dropped a hand on my shoulder. 'And now, my son,' he said, 'having counselled you longer by far than your brusqueness warranted, I'll lay a penance on you; a penance you'll perform, wholeheartedly, before we go any further.'

He pointed to the door. 'Through there is an exceedingly angry man; a man who, as God has arranged the affair, also happens to be our chief benefactor.' He smiled again at the stricken expression on my face. 'We own nothing for ourselves,' he said, 'as you ought to be aware. The funds that established this House, the very ground it stands on, came to us largely through the Senator's gift. You'll make your peace with him as best you can, begging forgiveness for your rudeness. Go on, I'll hear no more argument; a little humility won't do you any harm.'

The refectory was empty when I stepped back through the door; the last of the Brothers had gone about their business. Only Prudentius remained. He stood before the windows, back turned to me, hands gripped behind him. He stiffened slightly when I started to speak, but didn't turn. I gritted my teeth, swallowed, and made my apology to his unresponsive neck. I asked his pardon for my harshness, which had proceeded from lack of understanding; I pleaded my youth and ignorance, and begged him in his charity to end the enmity between us. He heard me through impassively before he turned; then I saw to my amazement that he was smiling broadly. 'My son,' he said, 'I forgive you with all my heart. Your apology was handsomely made. For my part I believe God speaks in you, though curiously; and I myself have known . . . ahem . . . older heads than yours turned by excess of zeal. Consider the thing forgotten; while with regard to your request, I promise I'll confer with the Father here to see what can best be done. As you realise, I'm not entirely without means. But on one condition, mark you, one condition only.' He wagged a thick forefinger sternly under my nose. 'You'll show me, before I leave this place, a repentant heretic. The Scythian must be brought to the Light; and through him the souls of those wretches he has so recklessly imperilled. . . .'

So the thing was ended; and shortly afterwards I took my leave of that pompous, untalented, generous-hearted man.

'So, Ulfilas,' I said, 'there the matter stands. You've been leaving this camp without my knowledge or permission; worse, you've been stirring up trouble and unpleasantness in Massilia, which is a poor return for all I've tried to do for you. Now you'll make recompense. You'll own the Son as you own the Father, unreservedly; and you'll do it at once, before the Bishop of Massilia. You'll be formally baptised into the true Faith; then you'll come with me to Cassianus and his sponsor to vouch for it in person. Is that clear?'

He didn't answer; just stood in front of me with those glowing eyes of his, hands clenched rigidly at his sides.

'If you don't,' I went on equably, 'you'll find life growing somewhat tedious again. Don't imagine for a moment I shan't punish you; also don't entertain any illusions about the form my spite will take. I've no intention of giving you the satisfaction of becoming a public martyr; you'll simply find yourself back where you came from, tethered securely in a place from which you'll have no opportunity for the conversion of the unsuspecting to any brand of Christianity at all. For the rest, temper your stubbornness with a new and beautiful thought. If Jesus was truly the Son of the Highest, then the State, far from merely denying your God, went to the length of crucifying him in person. For you, that should open brand-new vistas of loathing. Now get out, and find yourself a respectable tunic. You leave in half an hour; in one direction or another. . . .'

I had gauged him accurately. The light of Heaven might have meant little enough to his sullenness; but to be penned, after his freedom, far from the golden opportunities of oratory and subversion, was more than he could stomach. The ceremony went without a hitch; and I afterwards repaired to Paeonia's home congratulating myself on a piece of work effectively, if crudely, done.

Cassianus' reaction to my visit was unexpectedly swift. A week after Ulfilas' conversion a heavy cart jounced its way into the mine compound. The Father had come in person to inspect the workings, bringing with him three or four of his Order who were apparently skilled masons and carpenters. What they saw left them in no doubt of the urgency of the job. A few days later a small convoy rolled to a halt in front of the Praetorium. Each cart was loaded with building materials, stone, bricks and balks of timber; the parting gift of the Senator, who I was told was already on his way back to his Tarraconensian estate. Work began at once under the direction of the Brothers, who I discovered had nearly all been artisans of one sort or another before taking their vows of poverty and obedience. A chapel was consecrated alongside the building site, and a new air of optimism began slowly but surely to pervade the whole establishment. Winter as usual slowed the work considerably, but by early spring of the following year a whole range of neat new huts stood ready to receive their inmates. Small parties had already taken up residence, but I was destined never to see the full fruit of my labours. Easter passed, celebrated with more than usual fervour in the camp, before Paeonius at last sprang his carefully baited trap.

Winter had likewise put a temporary end to my jaunts along the coast. With the first fine weather, though, the carruca set out once more, with myself and an excited Paeonia on board. Gildo, Paeonius' doorkeeper, drove us; apart from him we were for once unattended. The Libyan had changed his usual robes for the costume of a charioteer. He was a surly brute, who I'd never heard exchange more than the odd grunted word with anybody; but he certainly looked spectacular enough, and his handling of the horses was excellent. We made fast time to the town gates, followed the Via Domitia up to and beyond the mine.

Paeonia had taken a fancy to see the mysterious pipeline she'd heard so much about, through which water still flowed defiantly uphill in contradiction of the laws of God and man. I'd warned her there wasn't much to look at, but she wouldn't be deterred. As it happened, the idea suited my own plans nicely. There had been no time to spare for routine maintenance on the channel; I had been intending to ride up and inspect it anyway at the first opportunity. Once beyond the camp we left the highway, turning on to the track that wound up through the hills. As the aqueduct came into sight I called to Gildo to follow its line as closely as possible, and to drive slowly. I leaned over the side of the carriage, noting the places where growths of moss round the slabs indicated leaks that would have to be plugged. I marked some of the worst spots with stakes driven into the earth. It was past midday before we reached the little valley at the head of the lake.

The appearance of the pipeline was as unimpressive as I had promised. Rank grass and weeds had seeded themselves round the cisterns; I prized the cover from the nearer tank to show Paeonia the clear water welling up inside. Afterwards we followed the course of the gully westward, leaving Gildo to attend to the horses. Some half mile farther on, the game path we were treading curved away from the valley, circled a hollow screened by bushes and tall saplings, hazed with fresh green. It was a delightful spot, one that promised shelter from the still-keen wind. Paeonia ran ahead of me into the little dell, calling excitedly; I followed more carefully, clutching the basket we had brought with us and that contained our meal. We spread a cloth on a rock and ate, sharing a flask of light wine; afterwards we sat chatting idly, at peace with ourselves and the world. An hour maybe passed, and I was wondering vaguely what had happened to the Libyan when I felt Paeonia's fingers tighten convulsively on my arm. I stiffened in turn, following the direction of her gaze.

What I saw sent me icy cold. Halfway up the opposite slope, black against the golden shimmer of sunlight, stood a massive wild boar. He must have turned aside from the path—I remembered, belatedly, the danger of following such tracks in unknown country—to grub among the roots and bushes of the hollow. He had come silently, upwind of us, and hadn't scented us; but he had certainly seen us now. He was studying us with one glistening eye; I saw him toss his head, scrape the ground with a neat, wicked little hoof.

There was nothing I could do. I was unarmed, except for the short dagger I usually carried; not that any hand weapon would be of the slightest use if he chose to attack. None of the saplings were stout enough to climb, though even if they had been we would never have reached them in time. He was barely twenty paces from us; if he decided to charge he would cover that distance before I could draw a breath. From what experience I had had of the creatures I knew they were generally shy, avoiding mankind wherever possible; but the knowledge was little comfort. This brute was old; I could see the worn patches on his hide, the criss-crossing of scars on his massive shoulders and neck. A solitary male, if I was any judge, driven from his herd by infirmity or age, and wholly unpredictable.

Our best chance lay in immobility; I whispered to Paeonia not to move, and to keep her eyes on the ground. The minute or so that followed seemed like an age. The creature stayed stock-still; once he stamped again uncertainly, and snorted. My heart came into my mouth; then, with dramatic suddenness, he wheeled, crashed back through the bushes to the lip of the dell. His hoof-beats drummed on the hard ground beyond, faded into silence. Paeonia made a sound halfway between a cough and a sob, and flung herself into my arms.

What happened next was as unexpected as the intrusion of the boar. I soothed her, stroking her hair, holding her close to stop the trembling; and suddenly it seemed her nearness, the warmth and firmness of her body, compounded with gross relief to work a change in me. My lips closed on hers, nearly involuntarily. She stiffened; then she was moaning, deep in her throat, jamming her tongue and teeth against mine. We rolled from the bank on which we'd been sitting, into a crackling tangle of branches. Her hair fell across my face, I reached for her again; and she was pushing at me desperately, thrusting me away. I saw her eyes, wide and appalled, the shocked circle of her mouth; then she was up and away, scrambling across the sloping ground. I would have caught her but a boulder turned under my foot, flinging me heavily backwards. I struggled up and saw for the first time, framed in moving leaves, the grinning face of the Libyan.

I panted to the top of the slope. By the time I reached it there was no sign either of Gildo or Paeonia. I began to run, calling her name confusedly. In my haste and distress I missed the path. I found it eventually, hurried along it. I was rewarded by a glimpse of the carruca dwindling in the distance.

I sat on a rock and cursed every God I could bring to mind and most of my acquaintances, not forgetting myself and the Libyan. When the fit passed I started to walk, hoping at each bend of the road that I would see the carriage waiting for me.

There was no sign of it. Evening found me still walking, but by that time my grief had changed to fury. Had I known then what I later guessed, that after the first blind panic she pleaded with Gildo, begging him to turn, things might have worked out differently. But the grim doorkeeper had his orders; he held the carruca at breakneck pace, not checking his lathered team till he dragged the horses to a halt outside his master's door.

Night had fallen when I met an oxcart, driven by a bad-tempered peasant. A couple of pieces of silver finally persuaded him to turn back. Undignified though the conveyance was, I climbed aboard it gratefully. I had no cloak, and the wind had chilled me to the bone; also the memory of the encounter with the boar was still too fresh for comfort. I finally plodded into the mine compound at midnight. I took myself to my quarters, and for the first time in many months spent a couple of hours studiously getting drunk.

I passed an appalling night. I did my best to keep the rage at fever pitch, but it was useless. All that remained was the memory of Paeonia. I saw the sunlight sparking in her hair, the close golden texture of her skin; saw her, in fact, with greater clarity than if she had been beside me in the flesh. I groaned hopelessly, cursing in the dark. I had declared myself for what I was, a hopeless and importunate lover; now nothing would ever be the same between us again.

I rose before dawn, hazy from lack of sleep, and made my way to the camp baths. An hour's sweating cleared my head a little, but I hadn't broken my fast before the rumble of wheels in the compound warned me I had a visitor. I walked to the door of the Praetorium. As I had suspected, the carruca was back; and with it a thoroughly enraged Paeonius.

Chapter twelve

He stamped ahead of me with his face like a thundercloud, refusing all offers of wine and food. 'Here's a fine thing,' he burst out when he had me on my own. 'Here's a fine thing indeed, a fine repayment for the trust I placed in you. What do you have to say for yourself? I hope you're satisfied with yesterday's work?'

I pointed out wearily that whatever the cause of the commotion there was certainly no need for it; and he drew himself to his full height, stared at me red-faced and quivering with rage. 'No need, sir?' he bellowed. 'No need? My daughter dishonoured, and no need to make a fuss! Yes, dishonoured; and take that look off your face, I know by who!'

I think my jaw must have momentarily sagged. Distressed though I was, I'd never heard such arrant rubbish in my life. What story he'd been fed I had no idea, but he was obviously labouring under a vast misconception. I tried to explain that all I'd done was comfort the child after a bad fright, that I wouldn't willingly have harmed a hair of her head; but he wasn't to be placated. He had a witness, he said, an unimpeachable witness; the slave Gildo, for twenty years his true and faithful servant, had seen the whole affair. The Libyan was prepared to swear, if necessary before the Bishop of Massilia himself, that the girl had been deflowered.

I sat back appalled. It seemed Paeonius took my silence for fresh proof of guilt; he proceeded, puffing, to embroider such a tale of devilry that I felt my own colour starting to rise. 'Very well, then,' I said, losing my temper in turn. 'I'll go on oath before the Bishop as well; and we'll see whose word stands, that of a Roman officer or that of a blackamoor. . . .'

He looked up quickly at that, with an odd expression in his close-set eyes. 'I wouldn't stretch your honour too far if I were you,' he said. 'I don't think in your circumstances it would be wise.' Something in his tone quietened me. I had the disagreeable sensation of meshes once more closing swiftly and silently round me; I stared at him, waiting for him to reach whatever proposition he had in mind.

'That's better,' he said. 'Yes, that's very much better. Now you're using your intelligence.' He was sweating profusely; he wiped his forehead, and it suddenly occurred to me that the rage and grief were mere play-acting. And pretty bad play-acting at that; Paeonius being a villain was if anything more absurd than Paeonius standing on his dignity.

'There's only one honourable answer,' he said, mopping himself again. 'And you know what that is as well as I do. You'll marry the girl, as quickly as possible, with all due rites and legal ceremonies . . .'

I couldn't help myself. I laughed outright at the absurdity of the whole thing. 'What?' I said. 'What, me? Without a penny to bless myself? How could I support her? What with, in God's name? Do you imagine I'm rich?'

'Don't try and hedge with me,' he said viciously. 'I know you're well enough off. I know a lot more about you than you seem to realise.' He waved his arm irritably at his surroundings. 'You even had cash to spare when you came here for bath-houses for slaves; if anybody ever heard of such a thing. Who pays out good money like that except a rich young fool, with more gold in his pockets than sense in his head? You answer me that. . . .'

The situation seemed to be going from bad to worse. I'd come to the conclusion a long time back that he was more cunning than intelligent, but I wouldn't have believed he could so far have allowed avarice to overcome his native wit. I opened my mouth to answer, but his next words stopped me dead. 'Don't think of refusing,' he said. 'Not if you're wise. Or it won't just be a matter for the town Praefect. The Emperor's going to be advised of a certain little affair in Burdigala a couple of years ago, that ended with the death of one of his officers. Don't bother to deny it,' he said, seeing my expression alter. 'I told you I had my contacts, and they're reliable, believe me. Don't think I don't know why you came riding down here, all good works and holier intentions; and don't think I don't know how the business was hushed up so effectively in the first place. Money again, my friend, money and influence. But it won't work twice for you, not if you get on the wrong side of me. You'll stand trial for murder, if I have to bring the action myself....'

I couldn't think of a thing to say. His last accusation at least was perfectly reasonable. My good luck was still wholly inexplicable; but faced with a similar case I'd have jumped to the same conclusions myself. It all fitted in very neatly; even my efforts at the mine could be explained as a desire to ingratiate myself with the local population, at the same time enhancing my reputation for virtuousness. I felt like laughing again, but if I'd once started I think I'd have been unable to stop. Already in Rome I'd suffered for my poverty; now it seemed I was be to victimised for my wholly imaginary wealth. Doubtless the spectabilis entertained some real measure of affection for his daughter; but equally he'd seen in my arrival an opportunity to revive the flagging fortunes of his house. It was straightforward blackmail; none the less it seemed he'd had to salve what passed for his conscience by trumping up a groundless charge of rape. That more than anything else was the measure of the man. I wondered distractedly how I'd ever imagined I liked him and how I'd come to be such an absolute bloody fool. If the business went no further than Massilia it could be unpleasant enough; I'd made no friends in the town, with the possible exception of Cassianus, while if the Libyan was prepared to perjure himself there were doubtless others just as ready to come forward with lies. Certainly Paeonia would be allowed no voice in her or my own defence. But it wouldn't stop in Massilia; that much was plain.

'Think about the advantages,' said Paeonius, still sweating. 'It isn't as if you're being asked to marry a harpy. You've always been attracted to the girl; she's a lovely creature, she'd make you a first-rate wife. As a matter of fact I've already got a property in mind that would make an ideal home for you both. Belongs to a friend of mine, but I'm sure we could come to a reasonable agreement. Better than sweating away in a mine office, eh? You wouldn't have to hide out any more, you'd be a respected citizen of Massilia. The Bishop's on your side, I know that for a fact; you'd find plenty of people ready to speak for you if it became necessary. And, after all, I'm still thinking first and foremost of Paeonia's well-being. Now then, what do you say? Be sensible, man . . .'

I clutched at a straw. 'You talk of your child's welfare,' I said bitterly. 'But you're not proposing much of a match for her, are you? Marrying her off to an unconvicted murderer? You're only presuming the affair's over and done with; what if I told you it wasn't?'

He twined his fingers, stared down at them, cracked a knuckle and smirked in a shamefaced sort of way. 'I knew we should come to that,' he said. 'I knew we'd come to that. Well, it's a point. It's fair. I admit it's fair. . . .' He looked back up at me under his brows. 'As things stand,' he said, 'I'm sure you'll agree with what I had in mind. I'd naturally expect you to make . . . ah . . . a settlement for the girl. A little . . . provision for the future. Enough to maintain her if, you know . . .'

He let his voice tail off. I could only stare at him, shaking my head. So I was to provide the dowry. I might have guessed as much.

I rubbed my face. Further argument was obviously useless. I needed time to think; not that I could see there being any way out of the impasse. 'Well,' I said finally, 'I'll consider what you've said. But as you'll appreciate, this is a serious matter for me. I can't just give you an answer here, on the spot.'

That seemed to satisfy him. He rose to leave, drawing his cloak round him. 'That's fair enough,' he said. 'That's fair. I'm not . . . don't wish to be . . . an unreasonable man. I only wish . . .' He seemed to realise he was weakening, and set his jaw. 'I'll give you three days,' he said. 'Three days from now. If your answer isn't in my hands by then, a courier will be on his way to Mediolanum. Goodbye, Tribune. Think carefully on what I've said.' He made as if to grip my shoulder, coughed, changed his mind, and waddled hastily from the room.

Under the circumstances I hadn't the heart to go through the farce of seeing him to his carriage.

My state of mind over the next two days can better be imagined than described. Quite obviously the match was an impossibility; equally certainly, I would never convince Paeonius of the true state of my finances. Wild schemes spun through my head. Could I, after all, find the money for a settlement somehow, make my home in Massilia with the girl I had courted so oddly? There was a certain dizzying appeal in the idea, but it was out of the question. The town was impoverished, I knew that well enough; but even if cash had been plentiful I had no collateral against which to raise a loan. And in any case—here was a new and highly unpleasant thought—how long would I remain at liberty after I met Paeonius' terms? What proof was there that he would keep his part of the bargain? It could be once the head money was in his hands I'd find myself delivered not to a bridal couch but to a squad of Imperial soldiers. It was difficult to believe the spectabilis would stoop that low; but then, he'd already stooped to blackmail. Even if he kept his word, though, the killing still wasn't answered for. If Paeonius had discovered the truth about me, others could just as readily. Time had dulled my apprehensiveness, but in no way lessened the danger; I could still be hauled away at any moment, made to stand hopeless trial for my life.

What if I threw myself on the mercy of Cassianus, begged sanctuary in the monastery? The Father might or might not be prepared to shelter me; but such a course would be an open admission of guilt, while I was still stubborn enough to find the notion of simple flight abhorrent. I thought bitterly of the load after load of silver I had consigned to the Imperial Mint. Had it not been for my appalling honesty I would by now be, if not rich, at least well on the way towards prosperity.

The more I worried over the mess, the more it seemed there was only one course left to me. I brooded over it by the hour, revolving all the factors in my mind. Eventually I reached my decision. On the second night after Paeonius' visit I set about getting my things together. I was sick, sick to the core; I could stand that sword of Damocles hanging over me no longer. I would ride to Mediolanum of my own free will, seek an audience with Honorius himself and suffer the judgement of the Emperor. The day I had been dreading had finally arrived.

The packing took longer than it need have done. I finished eventually, stood the heavy panniers against the wall. My dress uniform was laid out, ready for the morning; there was nothing else I could do. I stood rubbing my face, thinking about a cup of wine, and heard hoof-beats outside in the compound.

I walked to the door of the Praetorium with a vague feeling of impending disaster. The horseman was well muffled in a cloak, but the torches in their brackets gave enough light for me to recognise him as a servant of Paeonius' household. He handed me, silently, a parchment packet, rode away into the dark without waiting for an answer. I took the thing inside. One glance at the bold, sprawling handwriting of the address convinced me it wasn't from Paeonia. I broke the seals impatiently, stared at the signature. The letter came, of all people, from the Domina Papianilla; it instructed me, curtly, to ride to Massilia at once if I wanted to save my skin. I hesitated over it, pulling at my lip. The same presentiment, stronger than before, urged me to ignore it; but there was no real choice. I called Baudio, told him to saddle a horse. It was late already; I made the best time I could towards the town.

The streets of Massilia were deserted. It was past midnight before I reached Paeonius' mansion. I tapped cautiously at the side entrance. Nothing happened. I knocked again, louder, and at length heard the creak of withdrawn bolts. The door opened fractionally; the woman inside stepped back at sight of me, motioning me to enter. I walked through uneasily, heard the bolts shoot home. I followed the slave through the gloomy corridors of the place. She held a lamp high; in its bobbing light the statues once more seemed to watch me as I passed. I sensed again that pervading air of dampness and decay; to my overwrought imagination it was as if I was walking through the chambers of a tomb.

My guide paused finally before a doorway closed off by sombre drapes, tapped softly against the frame. A low voice answered her. She held the curtains apart for me, mutely; I ducked my head and stepped inside.

The bedchamber in which I found myself was sumptuously furnished. Paintings covered its walls; the light from the one hanging lamp was too dim to disclose their subject but I had an impression of a crowd of people, pale-faced and with upturned, beseeching hands. The figures seemed to shift and move at the corners of my sight. At the far edge of the little pool of brightness, reclining on the couch, was Papianilla. She had discarded the blonde wig; her hair, thin and greying, was drawn tightly across her skull, fell lank to her shoulders. In one hand she held a crystal goblet; as ever, the fingers that gripped it gleamed with rings. She was wearing a robe of dark, shadowy red; above it her white face seemed to swim in the gloom. She watched me for a moment before she spoke; then she laughed throatily. 'Approach, Tribune,' she said. 'I don't intend to eat you.'

I walked forward slowly into the light. Another wait, while she stared up at me mockingly; then she gestured with the cup. 'Be seated,' she said. 'Make yourself at home. Will you have some wine?'

I sat, stiffly, on an embroidered stool to one side of the bed. 'Thank you, Domina,' I said. 'Not at present.'

She shrugged, and poured me a glass anyway. In the quiet, the slop and tinkle of the liquid sounded loud. 'Here,' she said, holding the cup out unsteadily. 'Take it. Don't stand on your provincial dignity with me.' The rim of the goblet chattered momentarily against her teeth as she drank. She lowered it, and stared inscrutably again. 'What rubbish,' she asked, 'has my husband been talking to you?'

I told her, coldly, that he had seen fit to accuse me of rape, and had given me three days to consider marriage to Paeonia before placing in front of the Emperor certain charges that might or might not be substantiated. She heard me through, then flung her head back and laughed again. It wasn't a pleasant sound. 'The noble Paeonius,' she said, 'has no detectable principles whatsoever. Unfortunately he is also a fool; so most of his villainy is turned to no account, or recoils on his own head.' She drank again, grimacing; drained the cup, and instantly poured another. 'I've no doubt he also regaled you with the tragic history of my stepdaughter,' she said. 'It's an elegant tale, that suits the household's dignity better than the truth. Her mother was a pretty creature by all accounts; but unfortunately she suffered with a falling sickness, during an attack of which she struck her head and died. The child was born with that impediment; I doubt now that it will ever leave her. I think perhaps she has the sickness too.'

I rose. I had no intention of hearing Paeonia maligned, and said as much.

'Sit down,' she said. 'You'll hear what I choose to tell you, and you'll listen for as long as it suits my pleasure. Because you have no option.' She reached behind her. 'I've never been unfriendly towards you, Sergius,' she said. 'Or wished you ill. Come now, let's stop this silliness. Empty your glass, and have some more.'

'Thank you,' I said distinctly. 'No . . .'

'Then go without!' She raised her voice startlingly, flung the mixing bowl to the floor. It rolled, clanging; wine splashed darkly across the mosaic. The violence of the movement disarrayed her robe; I sat silently, averting my eyes from the pale bulge of a breast. She covered herself, smiling. 'So,' she said, 'I disgust you. You with your fine sensibilities, your poems and literature. But you have so very much to learn. You think I'm drunk, don't you?'

I didn't answer.

'Oh, yes you do,' she said. 'You think I'm drunk, and it offends you. Well, I am drunk. Because this is my house, and this is my roof. My own roof. I buried one husband, and sold myself to another to get it; and now I sit beneath it, I do as I choose. I at least am honest.'

She picked the glass up again, watching me over the rim. 'What are your feelings towards Paeonia?' she asked. 'Do you cloak them in decent, well-sounding words as well? Do you tell yourself you love her? Or do you admit what you feel, a mettlesome combination of pity and lust?'

I swallowed, and stayed silent.

'No,' she said. 'Naturally, you would love her. Well, I'll tell you what love is. It's a fumbling, in the dark. The hasty sliding of flesh within flesh. And afterwards ... it's nothing. A void, an emptiness. And we still have the rest of our lives to live.'

She fixed me with her pale eyes. 'Do I still nauseate you?' she said softly. 'Perhaps ... and yet I tell you this. This puppy-love, that sears you to the bone, will pass. You'll come to prefer your Ovid to your Propertius, for all your high ideals.' She smiled again, in the gloom. 'There, see,' she said. 'Papianilla discourses on poetry. Perhaps, in time, I could even learn to please you.'

There was a pause.

'As I remarked,' she went on levelly, 'my husband is a fool. Too great a fool, it seems, to recognise another. For my part, I'm not prepared to see the household saddled with an impecunious son-in-law. I'm well aware this tale of rape is nonsense; I heard what happened from the child's own lips, who is insufficiently aware of the world to lie. Neither, I'm persuaded, would a court be much impressed by that black ape of a doorkeeper, whose talents, though varied, lie in other regions than his head. But interesting though it might be, I've no intention of having my husband's ingenious little concoction put to the test. In that I can help you, if your martial pride will condescend to it. Certainly nobody else will.'

Suddenly I had heard enough. The heavy air of the chamber was making my head spin; I rose again, stood looking down at her. 'Thank you for your hospitality, Domina,' I said. 'But I don't need either your advice or your assistance. Tomorrow, at first light, I ride for Mediolanum. Since you put the choice so succinctly, I'll place my fate in God's hands rather than yours.'

I headed for the door, but she raised her voice behind me. 'Very well,' she said. 'Add cowardice to your stupidity if you wish; and see how far it gets you.'

I stopped, back turned to her.

'Yes, cowardice,' she said. I heard her rise and walk towards me. 'Though I've no doubt if you dredge the depths of our literary heritage you'll come up with a more comforting phrase. You're extremely good at that. And I'll tell you something else about yourself.' Her voice rose furiously again. 'You've looked at me, since you first entered this house, as no woman should be looked at ever. Why? What God-given right do you have to judge me? Do you think you're such a man?' She circled me slowly, staring. 'A man,' she said, 'would have taken from Paeonia what a man should have, a long time ago. Why do you come trailing back here day after day, picking and whimpering? Couldn't you find a better specimen than our little half-formed flower? Daren't you lift your aim to a woman with all her faculties?'

My memory of the next few moments is vague. It seems I raised my arm as if to strike her. She seized my wrist; a wrestling, and her mouth was jammed hard against mine. She was panting, groping for me with her disengaged hand.

Shock and outrage lent me extra strength. I wrenched away, flung her violently from me. She rolled across the bed, lay gasping and glaring up. It seemed she might rise, and attack me again; instead, she began to laugh. When she had finished she sat up, flicking her robe into place. To one side of the chamber was a dressing table, its top strewn with jars and bottles. She seated herself at it, back turned to me, unfastened her hair and began to draw the strands of it methodically through a comb. 'Get out,' she said calmly. 'Go and mumble a prayer, then play with yourself in bed. It's all you're fit for.'

I rubbed my face slowly. My hand was shaking; I stared at

it, trying to still the movement, heard my voice speak as if from a distance.

'What,' I said, 'do you intend to do?'

The hand that held the comb checked momentarily. I thought she wouldn't answer; then she glanced at me expressionlessly over her shoulder. 'Why should you be concerned?' she asked. 'You've made your decision. And very Roman, noble and right it sounds too.'

I waited.

'I shall reduce my husband's bargaining power,' she said. 'That's all. . . .'

I said woodenly, 'The girl must not be harmed.'

'She won't be harmed,' she said. 'At least not permanently. You have my word on that.' She set the comb down, rose and walked back to me. 'All I require of you,' she said, 'is that you absent yourself from your normal place of work for a day or so. Don't concern yourself any further with my husband's amusing little ultimatum; by tomorrow night it will have lost much of its potency. Oh, there is one last thing. Never, under any pretext, show your face here again; or I might be tempted to take a hand in your affairs. And I should be more effective, I promise you, than the noble Paeonius. . . .'

The servant was waiting to conduct me to the door. My head still buzzed with rage; it seemed I glided to it rather than walked. Once on the open road I forced the tired horse to a gallop; but what I was trying to run from, Papianilla, myself, or the anger I carried with me, I could never say.

The same anger roused me well before first light. I took four men, a brace of mules and a waggon and headed away from the camp. We reached the pipeline not long after dawn. I worked solidly through the day, hauling what lining slabs remained from the bottom of the ravine. I piled them into the cart and set off back along the line of the channel, offloading the material at those points where repairs were most urgently needed. The hard work kept me from thinking too long or too clearly about the events of the night before. It seemed the rage I felt reached out to encompass all womankind. I neither knew nor cared what form Papianilla's scheming would take; as for Paeonia, she must fend for herself. She had, after all, rejected me; now events were out of my hands, I could do no more to help her. I hauled and sweated, cursing the working party till I'm sure they were all as sick of my face as I was myself.

The day, that had begun hot, remained close, with that odd taste and tingle to the air that presages a storm. Sure enough, towards evening, when I finally downed tools and allowed the tents to be pitched, vast masses of coppery cloud were towering threateningly in the west. As darkness fell, the horizon became alive with the flash and leap of lightning. I sat moodily before the flap of my tent, sipping wine and watching the threatening sky. I stripped eventually and turned in, lay tossing restlessly in the thick dark. I was dog-tired, but sleep was far away; the sultriness alone would have prevented that.

The storm broke around midnight, with a crash and peal like the rolling of huge stones across the sky. Rain drummed and roared on the fabric above my head. I rose, poured myself a cup of wine and stood watching the wild display that sorted so exactly with my mood. The lightning was nearly continuous, a purple-white blazing that lit the hills, showed me the shapes of trees and bushes bowed under the downpour. The campfire was extinguished, with a hissing of steam; while the track beside which we had pitched the tents became a roaring gutter for the floodwater streaming from the higher ground. It was an hour or more before the storm finally blew itself away, passed muttering into the distance. The air was fresher, cooled by the rain; I slept at last, only to be plagued by a monstrous dream. What it was I could never remember, but I woke from it sweating, in the first grey light of dawn.

I sat up, frowning irritably; and the sound that had roused me came again. A scratching at the tent flap, and a hoarse voice calling.

'Sir . . . Tribune, sir . . .'

I rose, slung a blanket round myself, and lifted the flap. Air rushed in on me, full of that intense, sweet chill that only comes after a night of rain. I rubbed my eyes blearily. At first I couldn't place the man who stood outside; then I recognised him. It was the carter I'd set to watch over Paeonius' house when I first arrived from Burdigala. He waited now uncertainly, a scrawny, unkempt-looking Gaul with a mop of ragged greying hair. I growled at him, asking him what the Devil he wanted at that time of the morning. He fidgeted unhappily but stood his ground. There were, he said, many carriages in the streets of Massilia last night; he was sure the Tribune would be interested.

I swore at him, violently. What comings and goings there might have been in the town interested me not at all. He scratched his head, screwed his eyes up and tried again. Did not these carriages, he asked, converge on the mansion of the Lord Paeonius? For a purpose best undisclosed?

Obviously a piece of silver was the quickest way to be rid of him. I found one, grumbling, and warned him not to waste any more of my time with tittle-tattle that no longer concerned me. 'What were these carriages?' I asked sarcastically. 'And what was this purpose, that shunned the light of day?'

He told me.

What I heard had me first standing stock-still with shock, then grabbing hastily for clothes and weapons. It seemed one madness had left me to be replaced instantly by another. I slung my swordbelt over my shoulder, buckled a dagger to my side and set off wildly, careering down the stony track to the Via Domitia. But fast as I rode, fear paced ahead of me. Fear for Paeonia, certainly; but also fear of dark and the night, and a thing older than the stones of Rome. Other rites had been celebrated in Massilia than the rites of the Christos; rites the Emperors, pagan and God-fearing alike, had laboured for generations to suppress. Only the women adore him, it is said; the nameless one, who has known so many titles. Once he was called Dionysius, once Bacchus. Now he is Antichrist; for he can never die.

I know what happened to her. I learned a little; for the rest, a dream or vivid nightmare came to my aid.

They fetched her, from her room, and she was bathed. About the house was a whispering expectancy, a giggling anticipation of horror and delight. There were faster hearts beating, fingers that shook and fumbled as they unbraided and combed her hair. She tried to ask what they wanted, what they expected of her; but there was no answer. Her servants had become her mistresses.

She was anointed, with oils that stung and burned. The wind sighed over the house, heralding the storm, turning the courtyard torches to beards of flame; while the burning grew deeper, drew itself into a knot that blazed with pain and light. She was dazed, perhaps with grief; she reached for loincloth and breastband, but the clothes were twitched away. Instead they brought others for her; dark and wet-looking, shot with gleams like clotted blood, stinking with the animal stink of Tyrian dye. The robe, she saw, was divided to the waist, baring her flank and hip with every step. A cloak was draped round her, a heavy cloak with a hood; she muffled her face in it, whimpering at the pain between her thighs. Her heart pounded her ribs; she was led to the outer door, for she could no longer see. Already it seemed she walked above the earth, treading the wind.

A carriage was waiting for her, closed and black. Her feet were naked; she slipped, barking her ankle on the step. She felt the pain, but distantly, lost in the greater confusion of her body. The streets were deep and hot as a well; she clung to the woman beside her, rocking as the wheels found the stony ruts of the roadway. The noise, the jingling and hollow clatter, seemed now to recede, now to echo monstrously inside her skull. She screamed aloud, using my name; and the storm broke, with a blaze of light. The flashes lit the jerking backs of the horses, the high walls between which the vehicle careered; showed her too, ahead and close, other carriages, swaying at the same frenzied speed. She screamed again, but there was nobody to hear.

The wild ride ended finally. The carriage passed beneath a gateway, bounced and slewed on cobbles. She had left Massilia, that much she knew; but where in all that tumult she had been carried she had no idea. Her hands were beneath her robe. She thought she was bleeding, but the gliding smoothness was sweat.

Her shaking legs would barely support her weight. Torches streamed, sputtering in the deluge; the spots beat her face and shoulders like so many fists. The lightning blazed again; she saw a house front, blind-eyed and baleful, tendrils of some creeper that swayed and tossed. Water splashed tart against her ankles, but couldn't stop the fire. There was a doorway; she passed through it, felt the dripping cloak drawn from her.

Her feet were on slimy steps. Ahead were more torches; and a din of drums and chanting, the high excitement of many voices mixed with the rumbling of the storm. She moved forward, drawn irresistibly. The torches made a smoky cave of light. In it gleamed bodies, breasts and shoulders drenched with wine that streamed like thin blood. She saw the water of a pool or reservoir, sparkling, blackish green. Girls plunged into it with burning brands, rose with the flames unquenched. A madness came on her then so that she tore at her own robe, needing to join them. She felt it slip away; and for the first time a drum began to beat inside her head. A great drum, whose every stroke was silence.

It's bad to burst in on a Bishop before he's finished breaking his fast; worse to pour into his astonished ears nonsense about revels and night sacrifice. The good man coughed and choked, spluttering. What, the house of Paeonius? The Domina Papianilla? Surely I was mistaken. They were, after all, well known to him; both worthy, excellent citizens. And what was this about carriages, night carriages in the streets? Come now, my boy, come, *come* . . .

There was no help there. I ran for the door, evading a handful of His Eminence's slaves. The horse was still standing, lathered, where I had left it. The rabble tried to follow: I lost them in the confusion of the streets.

Paeonius' door wouldn't yield to my beating. I renewed the onslaught; and it opened with a suddenness that nearly sent me sprawling. The slave who had answered it took one look and bolted like a hare.

I lived a nightmare. In it I ran through room after room, full of the smell of incense and a stink like death. I shouted till the place rang, '*Paeonia* . . .' I was answered by silence.

I found her, finally. She was huddled on her couch, covers drawn up round her chin. I would have lifted her, but the blazing white of her face prevented me. She was smiling, maybe smiling, but crying too; it had been good of me to come at last. I leaned closer. She could whisper, if the Devil would give her back her tongue. Whisper, whisper . . . what did she say? It all made nonsense. She had . . . what was that, *eaten*? Eaten what, Paeonia, what did they *say* it was. . . . I couldn't understand this waywardness about the Sin of Chronos; till I pulled the covers back, and saw the glistening red.

I leaped back as if propelled by a powerful spring. I'd told her too much, it seemed, about honour in all my ramblings; she'd worked well, inside her elbow with a knife, left nothing to chance. It was understandable, the whole affair was understandable; yet it came to me, in that frozen instant, that she had no *right*. No right to thus perform pleasantries on her person, to lie as Calgaca had lain in blood.

The housefolk were gawking and rustling at the door. I burst between them, ran back through the house. The Domina, I saw, had also worked too well; while Paeonia, who undoubtedly had loved me, had now to die to recompense a Roman whore. By such means, perhaps, the whole wide world maintained its equilibrium, balanced as it was on the shoulders of a God.

Behind me they had raised a shout of murder. I quickened my pace, having no intention of dying misunderstood. In the atrium Paeonius himself ran at me, fumbling to draw a dagger from his pouch. He tripped and fell headlong, squealing with fright. I could have killed him where he lay; instead I leaped over him, flung my weight against the great bronze doors. They wouldn't give; I doubled back the way I had come, and there was a slave with a lifted sword. I cut upward to the armpit, it being his life or mine. He fell threshing. It gave the others pause.

The sunlight in the street struck with a physical shock. I stopped, dazzled, unaware for the moment of the mob closing round me. Some had run from the Bishop's house; others were strollers, attracted by the din, eager to assist at the apprehension of a criminal. They waved an assortment of improvised weapons: stakes, clubs, mattocks and axe-handles snatched up from the nearest shop counter. The first blows fell while I still stood dazed. I was aware, dimly, that I parried and thrust; then the press had fallen back. I retreated, breathing heavily, up the steps to the porch of the house. In the street a man lay writhing, hands gripped over his face.

The rest seemed for the moment unwilling to close. I retreated again, holding sword and dagger up level with my eyes. Behind me I heard the groan of bolts as the main doors were unfastened. Soon they'd be at my back. Then Gildo came running, with a pattering rush. He was naked save for a loincloth, and held a stabbing spear gripped short. The mob yelled, and boiled forward.

It seemed I once more acted without the conscious intervention of thought. Beside me the nearer of the ornamental chariots leaned against the house front. I braced my shoulders, put my foot against it and heaved. The thing swayed and toppled. The trace pole took the Libyan across the chest, bore him backwards; one wheel sprang from the axle, bounded into the crowd. The chariot crashed down the steps amid a confusion of legs and arms; then trumpets pealed deafeningly down the length of the street.

I stared round uncomprehendingly. As if by a miracle, the roadway had filled with armed and mounted men. They wore intricate breastplates, masked helmets that glinted and shone with gold. Cloaks of scarlet and yellow hung from their shoulders, pennants of the same gaudy hues fluttered from lance-tips; and at his side each man carried the spatha, the great broadsword of the Germans.

They had formed themselves into a rough semicircle, enclosing the mass of humanity in the roadway. Now the war horns blared again, close and raucous. Instantly, with terrible precision, the tips of the crescent swung inward, forming a cordon between me and my attackers. Steel screeched on leather, and the mob waited for nothing more. One man, bolder or more stupid than the rest, cut at the nearest rider with a dagger. A sword blade flashed, fell with a thud; he flung his arms up, vanished beneath the hooves. The rest took to their heels; within seconds the street was deserted, and silent save for the sobbing of the creature I had blinded.

I had seen no troops like these in all my life. I lowered my sword, feeling the shaking start. Another signal and the ranks wheeled once more. Through them a man came riding. He was bareheaded; his hair, blond, light and impeccably groomed, clung close to his well-shaped skull. He wore a chased and gilded breastplate; above it was a quilted, finely patterned surcoat. His horse, a superb grey, was caparisoned in flowing silk; to my dazed eyes, mount and rider seemed to shimmer like a flame.

He reined the animal at the foot of the steps. The chariot lay on its back on the pavement, its one remaining wheel still turning idly; beneath it protruded the shoulders and distorted face of Gildo. The rider glanced down briefly, then back to me; and I met the stare of the bluest, iciest eyes I have ever seen. 'Well,' he said, in a cold, perfectly modulated voice. 'It seems you have a talent for difficulties, Tribune. . . .'

Before he opened his mouth I knew him; there could be no mistaking.

This was Stilicho.

Chapter thirteen

The inner walls of the pavilion were of dark, billowing silk. Lamps burning perfumed oil cast a warm yellow light. At one end of the place a Syrian girl sang plaintively, plucking at some stringed instrument. At the other I lay on a cushioned Egyptian couch. Facing me was the greatest general in the world.

He wore a plain white tunic; the mellow light smoothed the hard lines of his face, making him look startlingly young. Between us a series of tables were heaped with delicacies. In the centre of the largest stood a bowl of fine Caecuban. Fresh snow, packed round it in sparkling heaps, had reduced its temperature to that of a mountain stream. A servant refilled my cup; I drank, swilling the stuff round my mouth. To me, it tasted like ditchwater.

'I prefer a paler wine,' said Stilicho. 'The dark vintages fuddle the brain, thicken and slow the blood. But, of course, I am not a Roman. I hope my choice is to your palate?'

I muttered something appropriate.

'I trust you'll forgive any minor discrepancies of the cuisine,' the Magister Militum went on sardonically. He glanced round the richly appointed tent. 'I try to travel with as little ostentation as possible,' he said. 'Such luxuries as I permit myself are necessary for the maintenance of my good name. Roma always loved the tangible evidence of power; and age has not softened her.'

He signalled briefly to a slave. The dish before him was removed, and another proffered. He ate thoughtfully before raising his brilliant eyes again. 'Such enquiries as I have made,' he said, 'exonerate you from direct blame in this morning's affair. In so far as you are culpable, I attribute your actions to the rashness of youth; a rashness you should by now have outgrown. Certainly sacrifices have been offered by night, which is contrary to the law; I have imposed a curfew on the town, and instructed its Bishop and Duovirs to make a full investigation. The spectabilis Paeonius I hold at fault; his lands and properties are forfeit to the State, and he and his household banished during the Emperor's pleasure.' He saw me about to interrupt, and raised his hand. 'Enough,' he said. 'This is my decision. I have no time and less inclination to pry too deeply into the causes of civil commotions.'

He took a sip of wine. 'You'll doubtless be relieved to hear,' he said, 'your potentially suicidal interference saved the life of the child Paeonia. I've placed her in the care of my physician; when she's sufficiently recovered I shall send her to the Court of Mediolanum. She'll be well looked after there.' He saw my face change, and forestalled me again. 'No,' he said, 'you may not see her. Her mind is disturbed; she doesn't know her own family, and certainly wouldn't recognise a random lover. For you, she was dead. Let her stay so.' His manner softened fractionally. 'She's a delicate creature,' he said. 'In time she may recover her awareness of the world; but in my country she would never be considered fit to share a man's bed. She'll never bear strong children, or make old bones herself. You'd be well advised to forget her.'

I was silent, trying to adjust to what I had heard. From beyond the open flap of the tent came the sounds of the camp settling for the night; the tramp of feet and jangle of harness, a harshly shouted order. I heard the noises dimly. It seemed a weight was lifted from me; but I was too drained by the confusion of the day to feel emotion. I had been given a guard to escort me to the camp, though whether as free man or prisoner still remained to be seen. Baudio gawped at the irruption of the Palatini; I picked my things up from where they still lay packed, walking like a man in a dream. My last visit was to the strong-room. A consignment was ready for collection; I stared a long time at the stack of dully shining ingots before turning away. I had taken enough from Paeonius; I didn't want his silver as well. I flung the keys to Baudio. He said, 'You must be mad.'

The Germans were waiting in the compound. I mounted, leading the packhorse on which I had stowed my gear. An armed Vandal ranged up silently to either side. At the gates a woman ran to me, one of the slaves. She gripped my knee and sobbed. I stared down frowning, unable to think what she could want. The party halted; then a trooper rode from the rear, seized her arm and flung her away. She fell into mud, lay moaning and clawing her hair. The hopeless sound of her wailing followed me down the road. It was my last, and fitting, memory of the place.

Stilicho recalled me to the present. 'I hope you realise,' he said, 'that you're a very lucky man. Had I not arrived when I did, that mob would certainly have hacked you to pieces without further questions. There's nobody quite as merciless as a Roman citizen with a legal excuse for violence; it's a trait of the civilised world I've never particularly admired.'

I found my voice finally. 'What brought you to Massilia in the first place, sir?' I asked him. 'Had you business in the town?'

He glanced at me keenly. 'Yes,' he said. 'With you. I asked for you at the mines. They told me you'd been seen riding for Massilia like a man possessed, but nobody could tell me why. Bearing in mind your past record, my curiosity was roused; I decided to follow you.'

He took his time before continuing.

'Three seasons ago,' he said finally, 'I happened to be in Mediolanum when an extraordinary despatch came through. As I knew the sender personally, I interested myself in the affair. It seemed a rising young Tribune—Duke Vidimer spoke highly if incoherently of you, and added an earnest though equally chaotic plea for clemency—had been unfortunate enough to kill a fellow officer in armed combat. The details were somewhat unclear; however, I was able to reconstruct the affair to my satisfaction by questioning the trooper who brought the report. As these things sometimes chance I found the victim of the brawl was also known to me, having served a short time on my staff.' He smiled, without particular humour. 'Under the circumstances I felt a certain kinship with the officer who had so abruptly terminated his career, and exercised a privilege peculiar to my position. The report, the only one made, was mislaid; as far as Mediolanum is concerned, the incident never took place. If and when you return to your Province you will undoubtedly have to face a civil charge, but that lies outside my jurisdiction or my interests.'

He stared at me with his disquieting eyes. 'Understand me clearly,' he said. 'I neither condone your behaviour nor excuse it; under ordinary circumstances you would certainly have been flogged and beheaded. My views in this as in other matters are dictated by practical necessity. The State is no longer so rich in manpower that it can afford to squander lives. You are an efficient officer, with the further advantage of a sound education. In short, you are of more use to me alive than dead; do I make myself plain?'

I sat staring blankly. The shock of the reprieve, after the long nightmare of waiting, left me incapable of coherent thought. He saw my difficulty and didn't leave me to struggle for long.

'Obviously,' he said, 'your life is no longer safe here. Blood was spilled this morning; I judge that you acted in self-defence, but the town will hardly see it in that light. So as far as Massilia is concerned I'm taking you away to stand trial for previous crimes against the State. In fact I'm sending you out of Gaul on a personal embassy, for reasons I'll presently explain. I'm giving you a small escort and the acting rank of Praefectus. But on my life, involve yourself in no more scrapes like the last; or I promise you, wherever I might be, I shan't rest until I've seen your head.' He gestured for the dishes to be cleared. 'How much,' he asked, 'do you know about Britannia?'

I answered stammeringly that I had no first-hand knowledge of the island, never having visited it, but that it had been my mother's homeland and that I had spent some time while in Burdigala studying its geography and history.

He nodded curtly. 'Excellent,' he said. 'I heard something of the sort from Vidimer.' He leaned back, eyeing me reflectively. 'You'll be leaving at first light,' he said. 'You'll make your way to Gesoriacum, where you'll find the headquarters of the British Fleet. I'll authorise you to take passage on the first available ship. You'll be carrying despatches which you'll deliver personally to the Vicarius, the Duke of the Britannias and the Count of the Saxon Shore. You'll place yourself at the disposal of these officers, taking what steps you can to secure the defences of the island. Your knowledge of the territory will be an asset; while never having visited it you've had no opportunity to form confusing loyalties.'

I stared at him again. As far as I was aware the defences of the Province were adequate. The old Twentieth had been cut apart at the Frigidus, but Britannia still disposed a considerable list of auxiliary regiments, both horse and foot, and there were holding garrisons formed from the Sixth and Second Legions both on the island and in Gaul. I said as much, and he nodded.

'Yes,' he said. 'I'm recalling them.'

The Syrian had finished her singing. She bowed quietly and slipped from the tent. A night-flying beetle boomed at the flame of a lamp and fell, kicking. In the silence, the little sound of its landing was clearly audible. I said, 'Then you're abandoning the Province.'

He had been watching the insect as it struggled to right itself. He said, 'I abandon nothing.'

He leaned forward, poured wine for himself, drank, savoured the taste and set the goblet down. 'Theodosius made me Regent,' he said, 'giving the whole Empire, both East and West, into my charge. Have you heard of Radagais?'

I admitted I had not.

'You will,' he said. 'At the moment he's gathering warriors north of the Danube. Greuthungi. At the last count he had twenty thousand men under arms. Soon, either next season or the season after, he'll cross in force. As and when he does he must be destroyed, if it takes every able-bodied man in the Empire to do it. Another horde is mustering beyond the Rhine; while Alaric of the Thervingi has been appointed a Magister Militum. This was the news that brought me east.'

For the first time his face showed signs of animation. 'Twice already,' he said, 'I've had that bloody old cattle thief pinned. Twice I could have had his head. Twice I've been ordered to release the Eastern comitatus, and watched him slip through the net. At the moment he's arming from Roman arsenals. When he's ready he'll move again. Into Italia. The throats he slits will be slit with Hispanian steel.'

'But why, sir?' I asked him. 'Why?'

He laughed, tossed the wine back at a gulp and poured more. 'Because,' he said, 'Arcadius Augustus would sooner see the West ground piecemeal than me in Illyricum, and his brother grown too strong. Since Frigidus he's been administering Dacia and Macedonia, which properly belong to the West, as part of his own territory. Now his brother's claiming them, and he refuses to give them up.' He drank again, deeply; then he spoke more quietly. 'The noble Arcadius,' he said, 'would certainly have strangled the equally noble Honorius in his cot, had his nurses ever allowed him to reach him. As babies they both destroyed the toys they found displeasing. Now they have a new toy. The Empire.'

I said quietly, 'Can you not stop Alaric, sir? By any means?'

'No,' he said, 'I cannot. To destroy him, now, would be to destroy his master; and I am bound by the oath I gave Theodosius on his death-bed, to guard the Empire and protect his sons. My word is neither given nor broken lightly; I leave that to my betters. Unless and until Arcadius comes to his senses, my hands are tied. I'm concentrating all available forces in Italia. From there I can watch both Alaric and the Danube. I shall intervene only as and when it becomes imperative.'

He rose quietly. 'I'm making the rounds of the camp,' he said, 'as is usually my custom. I'd advise you to get some sleep. You'll be making an early start tomorrow, and riding hard.'

I stood in my turn. I said impulsively, 'May I attend you, sir?'

He considered for a moment, and shrugged. He said, 'If you wish.' A slave came forward, bowing, with a cloak; he drew it round him and preceded me from the tent.

Fires were burning on the perimeter, and sentries had been posted at intervals. The Magister Militum had a word for each man; it seemed he spoke all their tongues as effortlessly as Latin. I followed a few paces behind him. The night was mild and fine, stars hanging low and lustrous in the clear sky. In the west a faint afterglow still lingered; southward were the humped outlines of the hills that guarded Massilia and the sea. He paused finally, stood muffled in the cloak, a darker shape against the night. 'Rome,' he said, 'conquered all things. Except the most important. One victory eluded her; she could never conquer herself.'

I waited.

'These were her terms to the world,' he said. ' "Live like us, and we'll build you towns. Learn our tongue and we'll give you schools. Obey our laws and we'll make you rich. *Go your own way, and we'll crucify you.* . . ." ' He half-turned to me. 'Perhaps,' he said, 'the world is deciding on its answer. While our little toy Emperors play with their soldiers, and bicker over their boundaries.' He was silent again, brooding. 'I warned Honorius,' he said, 'on the day of his accession. But he had his answer ready. "While the Rhine flows, Stilicho," he said, "we are safe." While the Rhine flows . . .'

I hesitated before putting the question that was in my mind. 'Why,' I said finally, 'does the Magister Militum devote his life to Rome's service? He speaks as if he hates her, and the things she stands for.'

He didn't answer for a moment, just stood staring into the dark. 'My father hated her,' he said. 'And his father before him. Her soldiers oppressed us with walls we couldn't scale; they dragged our young men off to slavery, scalped our girl-children to sell their hair for wigs. I learned her speech and her ways, made myself the master of her soldiers. So for a while I humbled her. Stilicho the barbarian owns the world.'

Round us the shrilling of insects was insistent and loud.

'I still support her,' he said. 'I stand guarantor for her pride, and her stupidity, and her sins. Because in our time, and our sons' time, maybe the sons of their sons, nothing better than her rule will come; and because my word is pledged. You are not a Roman; so I speak more freely of honour.'

A single trumpet sounded behind us in the camp and I thought for a moment I saw him shudder.

'Roma hasn't forgotten how to reward her servants,' he said. 'My time's short now, my very bones know it. The Sisters have spun the thread. But I tell you this. If I die, the Empire of the West dies with me. . . .'

He turned abruptly to grip my shoulder. 'The watch is changing,' he said. 'Get to your rest, I'll speak with you again in the morning. Good night to you, Praefect.'

I stepped back a pace. I said, 'Good night, my Lord.' I left him standing alone, head bowed on his chest, chin sunk forward in thought.

Quarters had been prepared for me. I lay sleepless a while, heard the watch change again before falling into a doze. I dreamed then I was Stilicho the Vandal, caught in a net and flung into a roaring sea.

He sent for me an hour before dawn. He had already broken his fast; he sat at a table littered with maps and papers, dictating to a pair of secretaries. He dismissed them when I arrived and spent some time going over the situation in Britannia in detail. The sky had brightened by the time he finished, and the camp was echoing with activity. Tents were being struck and stowed, baggage loaded, teams harnessed to their carts. I followed him from the Praetorium, was almost bowled over by a rush of Palatine lancers. Horses were saddled and waiting; I mounted, trotting behind him. My little command was already lined up; forty hulking men, splendidly mounted and armed. Riconus, their leader, a burly, blond-bearded villain, acknowledged me indifferently. 'They're Belgic Celts,' said the Magister Militum as we passed out of earshot. 'They're good fighters, but they're an awkward bunch to handle. However, they're all I can spare.' He reined his horse and sat waiting, watching the units of the comitatus form into their line of march.

It didn't take long. Standards were raised above the vexillations; and he called me to him again.

'Goodbye,' he said, 'and good luck. Make the best speed you can; and for your own sake don't betray the trust I've placed in you.' He wheeled his horse. 'Power is like poppy-juice,' he said enigmatically. 'A drug that blinds and dazes, and finally kills. Avoid it, Praefect. . . .'

I saluted; the war horns blared; minutes later the noise and colour were dwindling down the road. He rode stiffly, not looking back; the last I saw of him was the shimmer of his silken cloak. Then the baggage waggons had rumbled into the distance, and we were alone. I swallowed, and wheeled to stare at the men under my charge. 'Tighten your ranks,' I said curtly, 'and let's get moving. You look like a gaggle of Hispanian muleteers. . . .'

We made better time crossing Gaul than I had dared hope. Every day, as we hurried west, we met units and detachments moving east; comitatenses from Armorica and Tarraconensis, mounted German auxiliaries, shaggy, undisciplined foederati under their petty chiefs. It seemed the whole of Gaul was on the march, streaming towards the threatened heart of the Empire. The tramp of feet, the rumbling of waggon wheels and clop of hooves, filled every road in the Province. Mixed with the soldiery were flocks of camp-followers, quacks and mountebanks, jugglers and whores, women and children from uprooted families. Town after town was filled with them, the taverns overflowing; moving west was like swimming against some vast, relentless tide. A single traveller at that time could easily have found himself in trouble; but one and all steered clear of the grim, compact file of Palatini.

In Aquitania we came on a sight I had never seen before, and maybe now will never see again: a full Legion on the march. We sighted her Standards not long after dawn; she was VII Gemina, from Hispania. The noise that came from her was audible for miles, the dustcloud she trailed visible long after she had gone from sight. At her head, behind the Standards and the tossing banner of the Christos, rode German and Hispanian auxiliaries. Then came the main body of her infantry, ranged by century and cohort; I saw seamed, bearded faces of Cantabrians and Galicians, mouths that split in white grins as jests and insults were exchanged with Riconus and his Celts. Next came the vast baggage train, cart after cart piled with gear and provisions, crammed with women and dusty, wailing children. Dogs yelped and skirmished in and out among the wheels; muleteers shouted, swinging their whips above the long, lumbering teams of animals. I saw catapults and onagri, some still mounted on massive swivels, obviously torn bodily from some city wall; behind were more carts, and a mile-long tail of auxiliaries. These were all Germans, Alamanni and Burgundians. I reined to watch them, searching vainly for the Standards of the Arcadians. If they were present, they passed unseen in the confusion.

Bringing up the rear, behind the hospital waggons and a final detachment of legionaries, was a handful of foederati, Saxons by the look of them, their symbol a prancing white horse. Riconus spat ostentatiously as they passed. The column receded, its hum and rumble fading into distance while the dust settled slowly on the fields to either side, the sounds of birds reasserted themselves. Riconus swore, banged his shoulders, spat again and moved away; a few nights later we rode into Gesoriacum.

I found the port in considerable confusion. There had been heavy raiding a little farther up the coast; many galleys of the British Fleet were in harbour, but they were on standby alert and couldn't be detached. I had authority to claim passage, but not to commandeer. I got what I wanted eventually, by dint of a lot of shouting and by waving Stilicho's letter under the noses of all and sundry. A merchant vessel was leaving with a consignment of wine; I was informed we could travel as deck cargo or not at all. Riconus pulled a long face when I told him.

I didn't feel much happier myself, having been brought up with a healthy Roman distrust of all things nautical, but there was no help for it. The horses were hauled and cajoled aboard; no sooner had we embarked than we were told that due to gales and contrary winds we wouldn't, after all, be sailing that day, and the whole process had to be gone through again in reverse. I fumed and fretted, but there was nothing to be done. Finally the weather cleared. We re-embarked at night, by the uncertain glare of torches; it was a miracle none of the horses ended up in the harbour. We slipped from our berth sometime in the early morning, groped through the blackened port and felt the sway and heave of the open sea.

It was an unnerving experience. I could see nothing, hear nothing save the creak of spars and ropes, the unfamiliar cries of the seamen. Dawn brought a violent squall from the south. Rain drove in dull-grey curtains, obscuring all but our immediate surroundings; spray mixed with hail lashed across the decks with much of the violence of slingshots; the bulky vessel heaved and rolled abominably while we clung to whatever handholds we could find, striving to keep our footing on the slippery planks. I ended by being violently sick, but the Celts and their horses suffered worse. The animals screamed and plunged, frantic with terror. One lay down and died; another fell, smashing its foreleg. Riconus cut its throat and had it tumbled over the side; the sight and smell of blood sent the others wilder than before. At one time it seemed we might put about; Riconus promptly swore he'd send the vessel's master the way of the horse rather than endure the whole nightmare again at some future date.

The violence of the wind increased. Now there was no question of turning back; we ran before it, expecting any moment the mast and straining sail to be wrenched bodily out of the ship. Water poured in streams across the deck; the noise of it sloshing among the stacked wine-jars in the hold added a new dimension of terror to our discomfort. I know I personally commended myself to my mother's blue-eyed Gods; then, as quickly as it had arisen, the storm passed, hissing over the sea to the north. The sky cleared, and I caught my breath.

On our left, stretching as far as I could see, marched a line of colossal white cliffs. Headland after headland jutted proudly, seamed and cracked and bulging, receding each behind the next into a haze of distance. At their feet was the boil and spume of the sea; crowning them, grey and high and clear, were downs of wind-smoothed grass. No sight could have been more unexpected, or more typical of Britannia; for she above all others is a Goddess who loves to hide her face. She will veil herself for weeks and months with cloud and weeping mist, till you're sick and tired and wish yourself in Gaul or Hispania or blazing Africa, anywhere for a mere sight of the sun; then suddenly she smiles, and her spangled fields are green, and she is fair beyond the telling.

We had not been alone in the storm. A few lengths from our beam, dipping and rising in the heavy swell, lay a low, sleek galley, a scout ship of the British Fleet. Her sails, even the cordage that supported them, were dyed a soft, deep blue, the better I suppose to blend with the ocean; her crew wore uniforms of the same pleasing hue. She inspected us for some time before coming about smartly to cross beneath our stern. It seemed she was reassured as to our harmlessness. Orders rang, clear across the water; her long sweeps rose in unison, sparkled and dipped. She turned on her heel, trailing a wake of white-green foam; a very few minutes and I had lost her in the bright southern haze.

An hour saw us close in to the land; another hour and the cliffs themselves grew lower. We had intended to dock at Dubris, where there is a strong fort and a garrison, but the storm had driven us some miles from our course; I learned we would now put in to the smaller port of Rutupiae. It lay in a sheltered position at the head of an estuary flanked by shelving mudflats. As we neared the harbour entrance a heavy galley came threshing up astern; we stood well clear to give her sea-room. She had evidently been in battle. In her bow, above the wicked ram, was a gaping hole; farther aft were long streaks of what looked like dried blood. But her drum was beating, giving time to her crew, the banks of oars lifting and falling smoothly. She passed close enough for me to make out a bunch of prisoners on deck; big, powerful-looking men, loaded with chains and with pale hair and beards. I wondered to what part of the Empire they would be consigned, and what their end would be.

Rutupiae itself was congested with traffic, craft of all types tied up at the long stone jetties. We nosed our way eventually to a vacant berth; ropes were thrown and secured, gangplanks slung ashore. The sickness had passed, but I was still glad to set foot on firm ground again. I found the port in even worse turmoil than Gesoriacum. The narrow quay, backed by a line of dilapidated warehouses, was piled with produce, kegs and wine-jars, crates and boxes, barrels of oysters, balks of timber; even a cage containing a live bear. Everywhere the place was crowded with people. Some milled aimlessly; others, new arrivals it seemed, staggered along bent double under the weight of enormous packs. There were carts filled with children, women, poultry, geese, goats; all cackled, brayed, honked or yelped according to their natures, while underfoot were the inevitable scores of dogs. Port officials, bellowing themselves hoarse in attempts to instil some sort of order, merely succeeded in adding to the uproar. It was as if an entire town had uprooted itself and migrated for no apparent reason to the docks. I stared round confusedly. The Province, quite certainly, had gone raving mad overnight.

A tug at my sleeve made me turn. The lad who had elbowed his way through the crush was tall and well made. Brown curling hair framed a handsome, straight-nosed face; he was dressed, smartly enough, in the uniform of a military Tribune. He saluted, clicking his heels and grinning broadly. 'Sextus Valerius Nuadarius,' he shouted over the din. 'Office of the Vicarius. Welcome to Britannia, sir. . . .'

My arrival had been anticipated, then; though how the news had reached ahead of me I had no idea. I introduced myself, formally. 'Who's the senior officer here, Tribune?' I asked him. 'I need remounts urgently; I have despatches for the Dux Britanniarum, the Comes and your own office.'

'That'll be the harbourmaster,' he said. 'He's expecting you too, sir. If you'll follow me . . .'

I glanced behind me. Riconus and his men had succeeded, by main force, in clearing a space for themselves; the first dozen horses were already ashore. I could be of no help there; I followed Valerius, already shouldering a path back through the mob.

The office of the Port Praefect occupied the upper floor of a dingy building overlooking the wharves. A narrow stairway led to it; the Tribune tapped at a door, opened it and ushered me ahead. 'The Praefect Julius Constantius,' he said, rapping the words out like a German on parade. 'The Palatine Praefect Caius Sergius Paullus . . . *sir*!' He left the room with another rattling of heels, closing the door behind him.

The man who rose from the desk was middle-aged and prematurely lined, with grizzled iron-grey hair cropped close to the skull. 'Welcome, Praefect,' he said tiredly. 'Welcome to Britannia, come and take a seat. Had a good crossing?'

I told him the trip had been excellent apart from the storm. He laughed at that. 'Young Valerius was at Dubris,' he said. 'He nearly broke his neck getting here ahead of you. Baienius' —raising his voice—'bring me in some wine. Man in here with a raging thirst.'

He looked, and sounded, half drunk already.

'Sorry about the mess,' he said. 'Nasty thing to come in to. Been at the office since yesterday morning. Tried to sort some sense out of it. What's it looking like in Gaul?'

I didn't feel like wasting half the morning in talk. I told him briefly that troop movements were taking place eastward and added a request for remounts, as I had to be in Augusta without delay.

'Yes, those despatches,' he said. 'What's in 'em that needs so much hurry?' A stout, unkempt-looking man waddled from an inner room, carrying a bowl and wine-cups. Constantius filled one, shoved it across to me. I told him, stiffly, that I had no idea what orders I was carrying, but he merely shrugged. 'Have it your own way,' he said. 'I can guess, anyway. If they're what I think you've had a wasted trip. Legio II has already marched.'

'What?' I said. 'Augusta? What's going on, has there been a rebellion? Who are those folk on the quay, refugees?'

He shook his head slowly, no expression on his face. Then he rose to stand staring down through the one small window of the place. The noise of the mob reached up faintly to the room. 'Where are your eyes, Praefect?' he asked eventually. '*That is Augusta. . . .*'

Chapter fourteen

The road to Londinium was straight, featureless and dull. The first day's stage took us as far as Durovernum; there we met another fragment of the errant Legion, and turned aside for the night. Next morning brought a soaking, penetrating drizzle. Grey clouds rolled low overhead; flat saltings lay to our right, on our left was higher ground set with chequer patterns of square green fields. A chilling wind droned in from the marshes. The Celts rode for the most part in dejected silence; only Riconus remained enthusiastic. 'This is good land,' he said, time and again. 'Good soil. A man could raise good crops here. . . .'

What villages we passed looked ramshackle and untidy, collections of circular, conical-roofed huts surrounded by more or less strongly built palisades of pointed stakes. Valerius, riding at my elbow, indicated one such settlement scornfully. 'Look at that,' he said. 'Degenerates, all of them. They'll be scratching the ground with pointed sticks next. Name of the Gods . . .'

I glanced at him curiously. His dark hair was plastered to his skull, his cloak and leggings soaked, but like Riconus his spirits seemed unaffected by the downpour. I said, 'You're not a Christian, Tribune?'

'What, me?' he said scornfully. 'No fear, sir. My parents worshipped Jupiter and Fortune, like honest Romans. I follow Nodens, the Great Hunter.'

I hadn't heard of his godling. I forbore to say so; and he prattled on unconcerned. 'It's high time we had a real soldier to sort things out,' he said. 'If you don't mind my saying so. Have you ever met the Emperor, sir?'

'No,' I said drily. 'I haven't had that pleasure.'

'But you know Stilicho, of course. I think he's a great man. He did a lot for the country while he was here. I expect you've been to Rome, sir?'

'Yes,' I said. 'I've been to Rome.'

'I shall make the trip one day,' he said. 'It's my greatest ambition. But I feel my first duty's to Britannia. The Province is in a mess, and it's going to get worse. There isn't a decent officer between here and the Wall, except yourself, sir. . . .'

'Tribune,' I said, 'if you intend to become a good officer yourself, there's one lesson you'll take to heart right now. To keep a still tongue, particularly when it comes to your superiors.'

His face fell. 'Yes, sir,' he said. 'Very sorry, sir. It won't happen again.' He reined back, riding a pace or so behind me; but he looked so woebegone I soon called him forward once more. 'Tell me about these officers,' I said. 'But carefully. Carefully . . .'

He swallowed, and considered. 'Well, sir,' he said, 'the Count is a bar—a German, sir. Burgundian. He disposes infantry and cavalry, German and Hispanian, and some foederati. Forts at Portus Adurni, Dubris, Regulbium . . .' He reeled off a string of names, stretching as far as I could make out from the Sea of Vectis halfway to the Wall. 'He's a . . . competent soldier by all reports sir,' he said. 'But his men are thin on the ground already. What's going to happen if . . .?' He stopped, hastily; and I raised an eyebrow at him. 'Carry on, Tribune,' I said. 'The entire Province already seems to have made up its mind what despatches we're carrying. You're not alone.'

'No, sir,' he said. 'Thank you, sir. Well, if . . . troops were to be withdrawn it would make things almost impossible. We'd be relying on the Fleet, which is stationed at Bononia. It can't be everywhere.'

'No,' I said. 'I don't suppose it can.' I tried to draw my cloak tighter, to check the water trickling down inside my tunic. 'What about your Dux Britanniarum?' I asked. 'What sort of man is he?'

He hesitated again. 'He's a Tammonius, sir, from Calleva. They're a very old British family. He's—they say he's a good administrator, sir. Very popular with his men. They even call him Duke Marcus.'

I said curtly, 'But you don't think much of him as a soldier.'

He said, 'That's what's said about him, sir.'

'I see. And what is his exact command?'

'Oh, the north. Britannia Secunda and Valentia. Main forts at Eburacum, Luguvalium. And the Wall Forts of course. . . .' He treated me to another demonstration of his retentive memory. Loquacious he might be, but extremely well informed. Privately, I was pleased; what he had told me corresponded almost exactly with what I had learned from Stilicho. 'I see,' I said when he had finished. 'Well, you make Britannia sound as hollow as a nut. How do the towns of the south and west get by? What forces do they have?'

'They do the best they can,' he said slowly. 'They can most of them raise limitanei, of course, at a pinch. You get a few detachments of regulars at Imperial posting stations, places like that. But most of them spend more time on their allotments than drilling, sir. You'll see that for yourself soon enough.' He glanced at me sidelong, pulling at his lip with his teeth. 'I'd like to transfer to your command, sir,' he said. 'If you'd . . . would you consider it?'

I stared at him. 'I have no command, Tribune,' I said. 'I have the men you see behind us, that is all. I'm merely a messenger.'

He drew himself up in the saddle. 'Begging your pardon, sir,' he said stiffly. 'You're a Palatine officer. I'd like you to consider my request as formal, if you'll have me. I'll submit it in writing next time we stop. You'll need somebody who knows the country, sir. . . .'

We halted for the night at Durobrivae, stabling the horses alongside those of the small standing garrison. The room Valerius found for me was comfortable, but I lay and tossed restlessly, unable to sleep. Mine of military intelligence though he was, there was one thing the Tribune hadn't been able to tell me: exactly what was happening at Rutupiae. I was over the initial shock of seeing British troops; Stilicho had, after all, warned me the Legions of the Province had been down-graded, I might have expected what I found. But with Londinium barely over the horizon the problem still remained; what were the remnants of the Second doing at the port? A military detachment had been posted to the coast as part of the overall scheme for strengthening the eastern seaboard, that much I already knew. Why had they moved? On whose orders had they marched? Certainly not the orders of the Magiste Militum, the sealed packages still lay beside my couch. turned over irritably. There was one conclusion left, and that as plain as day: it was just that I didn't want to face it.

I'd left a lamp burning. I rose and mixed myself some wine. At least I'd avoided major involvement; I'd shown Rutupiae a clean pair of heels. My duty seemed plain: to deliver the despatches, and wait developments. For the present, I could do no more.

A faint sound at the door made me turn. I frowned, then walked silently to where I had left my swordbelt. I crossed to the door, gripped the catch and pulled. Valerius tumbled backwards across the threshold, sat up looking alarmed. He'd been keeping vigil outside the room, a drawn sword on his knees. 'Well, Tribune,' I said when he had risen to his feet. 'It seems neither of us can sleep tonight. So you'd better join me with a glass of wine.' I lit another lamp, picked up my tunic from where I had hung it across a chair. 'Valerius,' I said, when he was seated, 'tell me one thing. Who runs this Province?'

He licked his lips. 'Marcus Tammonius Vitalis,' he said finally.

'And?'

'Sir?'

'You were going to say something else. Out with it, man. . . .'

'He's a merchant,' he said unwillingly. And that, for the moment, was all I could get out of him.

Dawn brought at least one reprieve. The rain had stopped. We were on the road early. It descended steadily now, dropping into the great valley of the Tamesis. Marshes still stretched to our right; in the distance was the steely glint of the estuary. The wind blew keen from the sea, lifting our cloaks and the manes of the horses. We made good progress; it was midmorning before Valerius, riding as usual on my left, stiffened and stared ahead, checking his horse. A moment later I saw the riders too; the vanguard, it seemed, of a considerable column, heading towards us down the long straight way. We cantered to meet it; before long we were close enough to make out the insignia carried at its head. 'German horse,' said Valerius excitedly. 'Hispanian horse and foot. . . .' Then, suddenly, 'Legionary Standards . . . Sir! Sir! It's the whole Army of the North!'

'Riconus,' I shouted. 'Get your men off the road. Tribune, find their commanding officer. If it's Tammonius, tell him we're carrying urgent despatches from the Magister Militum. Look lively, man . . .'

The ponderous column, its tail still a mile away in Vagniacae, was halted. All along the high-shouldered road the clusters of legionaries and cavalry, the drivers of the carts and baggage waggons, lounged or squatted at ease. A pavilion had been hastily erected, on level ground at the foot of the mound; a gaudy, fragile thing replete with banners of silk. Outside it, grounded in the turf, were the Standards of the Commander-in-Chief. Inside, Tammonius Vitalis faced me across a trestle table, Stilicho's despatches in his hand.

He was a slightly built, dark-haired man with dark, worried-looking eyes. He wore the uniform of a Legionary commander. His helmet he had placed beside him on the table; his cloak had been slung carelessly to one side. He read swiftly, scanning the lines to the signature; I watched the furrows forming over the bridge of his nose. Finally he flung the thing down. He said, 'I can't do it. He can't ask this of us. Not this as well.'

I stood silently to attention. He hadn't asked me to sit. He said, 'I can't do it, I tell you. I can't.'

I said, 'I'm merely a courier, sir. I'm sorry to bring bad news.'

He said, 'Damn it, will you listen?' He had a smooth, nearly boyish face; at first I'd thought him younger than myself, till I saw the peppering of white at his temples. Now he looked hunted. 'There are ten thousand blood-drinking savages beyond the Wall,' he said, 'poised to sweep through the Province with fire and sword. In the west, Hivernia is armed from one end to the other. Niall of the Nine Hostages has ravaged the coasts nearly to Vectis, there are Scoti on the mainland beyond the Sabrina. The Eastern Sea is thick with Saxon war boats. Now we are to give our troops to Rome.'

He reached shakily for a goblet, scooped himself up some wine. 'This Magister Militum, this Stilicho,' he said. 'Who is he but a barbarian? Why did he come to Britannia? He spied on us. He sounded us, and found us rich. Now he's sold us. To his friends across the Rhine.'

'The Magister Militum,' I said coldly, 'is an honourable man.'

'A Roman talking of honour,' he said bitterly. 'God, what next!'

I opened my mouth to answer him, and he shouted me down. 'Be silent,' he said. 'I gave you no leave to speak. You are a courier, remember your words. I am the Dux Britanniarum. I control the forces of the Province. I, and no other.'

'You did control the forces of the Province,' I said, finally stung into retaliation.

'What?'

'I said you did. . . .'

'And damn your insolence, I say I do!'

'Your forces, sir, have rebelled. . . .'

'*They have not rebelled!*' He brought his fist crashing down on the table. He was still holding the wine-cup. It shattered; splinters flew, and he flung the thing away. He sat gripping his wrist; blood coursed brightly across his palm.

There was a circle of spectators at the open flap of the tent, drawn by the sound of raised voices. He stared round them, slowly, and shook his head. 'Leave us,' he said. 'Stand apart. Leave us alone.'

A splash of blood dropped on to his knee. He seemed unaware of it. 'I am . . . subject to pressures,' he said. 'And have been subject to pressures, greater perhaps than you . . . understand. My . . . office has proved a difficult one, hard for one man adequately to fulfil.' He looked up at me with a species of mute appeal. 'The Augustan garrisons have marched,' he said. 'I rode to intercept. As yet I . . . don't understand. It seems we are surrounded by traitors.' He leaned back wearily. 'Read me the despatch,' he said. 'Read it again.'

I read, quietly. When I had finished he nodded, eyes closed. 'Yes,' he said, 'I see. I understand. There has been no rebellion. There will be no rebellion. All units listed will embark for Gaul. I will be seen to have done my duty.'

There was a silence that I chose to break. I said, 'What are the Duke's orders in respect of me?'

'My orders? In respect of you?'

'The Magister Militum,' I said, 'instructed me to place myself at your disposal after delivery of the despatch.'

'Yes,' he said. 'Of course. The Magister Militum denudes the Province, but he sends us the reinforcements we begged for.

A Palatine officer, and a numerus of cavalry.' He stared round him vaguely. 'Do you see a cloth?' he asked. 'As you see, I have cut my hand.'

There was a side table, set with a meal. I took a napkin, passed it to him silently.

'Thank you,' he said. 'Would you help yourself to wine? Perhaps you would pour some for me too. There are . . . cups behind you, on the table.'

I did as I was asked. When he lifted the goblet his arm wobbled violently. 'I shall . . . camp,' he said. 'Outside Rutupiae. You will take your men south, to Portus Adurni. You will deliver your despatches to Hnaufridus, Count of the Saxon Shore. Yes, I must . . . despatches . . . the Fleet Praefect must be made . . . aware. And perhaps too we shall commandeer . . . You will write to Mediolanum, stating that in all respects I have—complied—with orders received. There has been no . . . rebellion. Do you . . . understand?'

I said, 'It will be as the Duke wishes.'

'Yes,' he said. 'Yes. Thank you. I'm sorry, Praefect, I . . . this is all . . . distressing to me.' He set the cup down, stared at the table. 'Meet me in Augusta in two weeks,' he said. 'The Provincial Council will be in session. I shall explain the . . . steps we have taken to secure the country's safety. And may God help us. God help us all. . . .' He peered at the red-stained cloth round his hand, pressing his finger against the palm. 'Mine was the first blood shed,' he said. 'The first British blood, for Rome. Perhaps it will be remembered of me.'

I saluted, stepping to the tent flap. I turned away quickly, but not before I had seen a curious thing. Marcus Tammonius Vitalis, Duke of the Britannias, was in tears.

I rode south grimly, by long stages. Valerius for once left me alone, warned, I think, by my expression. We travelled across country under the Tribune's guidance, eventually striking the main road to Noviomagus. A short further journey brought us to our destination. Despite my bitter mood I was impressed by the appearance of the fort. None of these British towns and camps seemed to have suffered the attrition that was commonplace in Gaul; everywhere I saw high, well-kept walls, over the tops of which loomed line after line of heavy catapults. If, as Valerius claimed, they all maintained limitanei of their own, I couldn't see any casual raiding party getting much satisfaction from them. The state of the country maybe wasn't so parlous as I had imagined, or Tammonius inferred. The rich villa estates between the towns presented quite another problem in defence. It seemed what was required was a series of mobile, well-armed units that could be moved rapidly to any threatened point. There were cataphracti, heavily mailed cavalry, in both the northern and southern commands, that much I knew; maybe they could be made nuclei of the new-style forces. It was something I must discuss both with Hnaufridus and the Duke.

Insensibly, I had worked myself into a better mood. My meeting with Hnaufridus further raised my spirits. The Count, a stocky, phlegmatic German, took my news, when it was translated to him, more stoically. Stilicho was likewise stripping his command, paring it to the bone; but the Burgundian merely shrugged. There was, he explained in a halting combination of Latin and German, a legend of a Day of Fire so consuming that the Gods themselves would be destroyed. If the deities included Roma, then that was well. For his own part he desired nothing better than to die with a sword in his hand, having first won general approval by the valour of his conduct and by killing many enemies. He showed a polite interest in my suggestions, though I had the notion he was more impressed by the gold torque I had been careful to wear than by what seemed to me at least well-reasoned arguments. I rested the Celts for a few days at the fortress before striking back towards Augusta, the administrative and trading capital of the Province.

I had heard a lot about the town. None of it prepared me for the reality. Outside Rome herself, it was the biggest city I had seen, but it seemed built to no discernible system or plan. Shops and tenements, warehouses, churches and disused temples jostled and crowded each other; there were mile on mile of dank, ill-smelling alleys and wharves, choked with filth and garbage, overshadowed by decrepit, leaning houses of wood and stone, the haunt of cats, curs, and naked, unscrubbed children. In every direction sprawling suburbs shoved and nestled against the massive walls; the houses had even crept out on to the one great bridge, till it seemed the timber piles could scarcely support the superincumbent weight of building.

Scattered at random, like ships riding a crooked sea of roofs, were vast public edifices. I made my way to the largest of them, the town basilica. I found the conference promised by Tammonius had been called for the following day. Most of the delegates, quinquennales from every major town of the Province, had already arrived; more were still crowding into Augusta. The town had been thrown into a tumult by news of the evacuation; it seemed I was in for a stormy session. I saw Riconus and his men settled in—empty stables, temples and warehouses all over the city had been pressed into service for the occasion—and asked Valerius to find us beds not too thick with lice, and not too distant from the great basilica. I was tired from the long days of riding, and would have preferred to sleep; but no sooner had we installed ourselves in moderately clean-looking quarters than a flamboyantly liveried escort arrived to conduct me to the presence of the Vicarius, already in session with Tammonius and the Praesides of Britannia Prima and Maxima Caesarensis, in which administrative division Augusta itself lay. The extraordinary conflation of civil and military authorities alone would have convinced me of the seriousness with which the Province viewed current developments.

The session took place at the private house of the Vicarius, a vast, rambling palace of a place a few streets from the basilica. It went on noisily and inconclusively well into the night. I took a back seat as far as was possible, content to listen and observe. By the time the meeting broke up, the officials with their trains of assistants making their several ways by litter and carriage to their lodgings, I at least understood what Valerius had meant by his veiled sneer about the merchant classes. It seemed they formed one of the most powerful and vociferous groups on the Province. One and all were terrified of the possibility of connections with Rome being severed; from the remarks of the elderly Vicarius—himself a prominent Augustan banker, and the owner of a considerable fleet—it seemed they stood solidly in favour of British intervention on the mainland in face of the growing threat from across the Rhine. From their point of view nothing but good could come of the evacuation; they were less concerned with the individual fate of the Province than the maintenance of trading links with Gaul. But a numerically stronger group, consisting mainly of the curiales and lesser landowners, was equally violently opposed to any reduction in British strength. Surprisingly, it was this latter view that Tammonius championed, in the face of dogged opposition.

He seemed to be fully in command of himself again; no reference was made to our unfortunate earlier meeting. 'You all know me, very well,' he said wryly. 'I've spent a small fortune already on behalf of the Province; I even gave up my privileges for the Army. Now I tell you this. We must protect ourselves; *we can't hold Gaul*.' Whatever his actual merits as a soldier, he was clear and level-headed in debate; and his loyalties were wholly British. Nobody asked me for my views, which was probably just as well. The noise, and the airless heat of the room, made my head spin; when I finally left it was with a grim sense of foreboding. A Council so badly divided against itself seemed a poor tool with which to soothe an angry and disgruntled populace.

Valerius roused me at some indeterminate hour before dawn. The first thing I became conscious of was the renewed pounding of rain on the roof. I had a headache, and a raging thirst; my throat felt as dry as on that long-ago morning in Belgica when I stormed the hill-top fort with Vidimer and his men. I dressed and shaved, working awkwardly by lamplight, forced myself to eat a little bread and fruit; then Valerius and I set out together through the slowly brightening streets towards the basilica.

Early as it was, the Forum and the ways that led to it were noisy with carriages and hurrying pedestrians. They splashed indifferently through the rubbish-choked swirls of the gutters, chattering to each other in the high-pitched British tongue. I had instructed Riconus to present himself with a dozen men. I found them waiting outside the basilica, muffled in damp cloaks, heavy-eyed and sullen after a night's carouse in whatever fleshpots the town possessed. Riconus it seemed had changed his mind about the Province; his muttered comments on towns in general, Augusta in particular and the grasping, untrustworthy nature of the British innkeeper did nothing to improve my temper.

We shouldered our way by main force into the body of the hall. It was immense, echoing and cheerless. Light from the big half-round windows to either side fell grey and livid on the mass of humanity that filled the floor. There was a reek of damp cloaks and tunics; while above the grumble of voices was the steady roar of rain on the long roof, the crash and splatter of water spouting from the eaves to the carved stone channels of the gutters.

On the rostrum at the far end of the place sat the dignitaries I had met the night before, with some new arrivals. Prominent among them were the Praeses of Britannia Secunda, a fussy, pompous little Roman, and the Bishop of Londinium, a Gallic nobleman of ponderous girth and even more ponderous dignity. The Vicarius, frail-looking and silver-haired, had robed himself for the occasion in a toga complete with massive purple stripe; the others, with the exception of Tammonius—once more in full military uniform—were dressed more familiarly in richly embroidered tunics. I deployed the Celts in a businesslike formation to either side of the platform, and mounted the steps with Valerius to join the Duke. Shortly afterwards some measure of quietness was achieved, and the proceedings opened with a short address from the Vicarius.

He told the assembly, somewhat circuitously, what it in fact already knew: that the Magister Militum, acting with the Emperor's full authority, had seen fit to withdraw from Britannia the regiments newly seconded to her defence, along with all remaining Legionary troops. Unrest was already evincing itself before he finished; when he sat down, the place burst into angry uproar. I looked down on a sea of shouting faces and waving fists. Riconus and his men, casting anxious glances behind them, began to edge towards the rostrum; the other guards moved forward, struggling ineffectually to quell the disturbance. Half a dozen individual skirmishes developed; the din was at its height when Tammonius rose, arms spread wide. The simplicity of the gesture achieved what no amount of bludgeoning seemed likely to do; the crowd grew quiet again, by slow degrees.

The Duke didn't begin to speak until the dim noise of the rain had once more reasserted itself. His speech was brief, blunt and to the point. He upbraided the audience for its sad lack of dignity; he reminded the curiales that their collective security depended on the continued strength of the West and stressed that Rome herself, the hub of the Empire, was in deadly peril. 'If Roma falls,' he said to the now-silent hall, 'which of her Provinces—Hispania, Gallia, Belgica, Africa—will survive? Will Britannia survive? For this is where our troops have gone; to Rome, to help her in her hour of greatest need. Rome first sent her Legions to us, in the time of our forefathers; are we, now, to grudge her our aid? What sort of people are we?'

There was a sudden commotion in the front ranks of the audience, down by the foot of the rostrum. An old man pushed himself forward, raising a long, gnarled staff. He was white-haired, thin to the point of gauntness; the eyes he turned up towards the platform were blank and blind.

'My eyes no longer see,' he said, 'but it seems I see clearer than any of you. What rights has Roma over us? You say she takes back her own. What right had she to take my son? Was he a Roman? Am I? Was my father, or his father before him, who also served her?'

There were angry shouts. A score of voices proclaimed heatedly that the speaker had no business in the hall. Hands were laid on him; and Tammonius stilled the mob again, with a furious gesture. 'Is this your courtesy?' he asked bitterly. When they were quiet he spoke directly to the old man. 'Who,' he asked, 'was your son?'

The ancient drew himself up. 'Caius Julius Thiumpus,' he said. 'Centurion, Third Cohort, Sixth Legion . . .' His voice faltered. 'One other son I had,' he said, 'but he was gathered to the Gods in the time of Theodosius. Now I have none to support me.'

The Duke spoke gently. 'Where have you travelled from, Father?'

'Luguvalium, by the Wall.'

'You have had a long, difficult journey,' said Tammonius. 'You did well to come. It grieves me that I can offer so little comfort. But I can pray, as we can all pray, for your son's safe return. For surely he will return, when the Legions return; in victory.'

There was a buzz of conversation.

'You there,' said Duke Marcus. 'And you. Find him a seat by the wall.'

The old man, still protesting, was led to one side. 'This is an agony of the people,' said Tammonius quietly, when peace was once more restored. 'We would do well to remember it. There can be few of us, in this hall, who have lost our sons to Rome. Let us think on this; for it is for the people's sake that we are here.'

He was right, of course. For over three hundred years Legions had been quartered in Britannia. Now their last pitiful remnants were gone. What had been left, what could have been left, that was Roman? Families had grown up, whole little dynasties, each generation with its tradition of service to the Standards; these last troops to cross the narrow sea had been Britannic through and through. Over half the world the same thing must have been happening; men had been torn from their families, from all they knew and understood, to fight in a cause they could barely comprehend, against an enemy they had never heard of, on behalf of a sprawling, bawdy city they had never seen.

The crowd was thoughtful; and Tammonius was quick to seize his chance. 'In the meantime,' he said, 'Rome has neither neglected nor forgotten us. For a little while it seems we must defend ourselves; and I call on her ambassador, the Palatine Praefect Caius Sergius Paullus, to outline the measures she wishes us to take. For our own good, for the Empire's good, and for the continuation of the everlasting peace.'

I stood up, sweating. This was something I hadn't expected. I had nothing prepared; I sent up a brief prayer to the memory of old Gellius, or to his shade if he no longer lived, and launched into an exposition. I repeated the main points made by the Duke, laying particular emphasis on the menace from the Danube; I reiterated my faith in Stilicho's honour and ability as a soldier; sketched in as well as I was able the golden tide of peace and prosperity the returning armies would bring with them, and ended by pointing out the immediate need for confidence and self-help. 'Your walls are strong,' I said. 'Such of your towns as I have seen are excellently defended. Hnaufridus, Count of the Saxon Shore, holds the south and east with an iron ring of fortresses; the narrow seas are safe; the Wall is manned; the noble Tammonius is in the north, between you and the barbarians. If to strong walls you add courage, patience and stout hearts, then Britannia will win the day. Enlarge your garrisons; see your young men are trained. In Rome, when she first grew to greatness, each citizen was a soldier at need. Let it be so with you; and the Emperor will honour you, speaking of you with pride.'

I sat down feeling well acquitted. Beside me Valerius was grinning broadly, but already a hubbub was rising. Fists waved again; and a man in a tunic of blue and yellow pushed his way to the front of the crowd. It seemed he carried some authority; they made way for him, and fell quiet to hear him speak.

'We have heard the words of the noble Sergius Paullus,' he said. 'We doubt neither his valour nor his wisdom. Yet I, Gnaeus Claudius Felix, quinquennale of Lindum, ask him this. Who among us does not remember the times of Magnus Maximus? Or the words of Maximus, when he first crossed to Gaul? For he too promised us greatness. And who does not remember Catena, the Chain with which we were lashed? Is not the arm of Rome . . .' A roar interrupted him, but he rode it down. 'Is not the arm of Rome long, her memory even longer?' He looked for the first time directly at me. 'We have paid the price for raising soldiers already,' he said. 'And that within the memory of our sons. The Roman Peace rules all; a man may not bear weapons; for generations this has been so. We hear, and we obey; we lay down our swords, in deference to Rome's will.' He raised his arms high. '*Then let Rome protect us. . . .*'

I'd started to sweat again. I'd known this was coming sooner or later. Technically, he was right; under Roman law, civilians anywhere in the Empire were expressly forbidden to wear arms except for the purpose of going on a journey; and Britannia had in no way been released from the stricture. What I had been led to advocate was not so far removed from treason; or so it could be construed by the handful of Imperial spies who were no doubt present in the hall. I began to see why Tammonius had been so keen to make me speak. I thought quickly, and rose.

At least they quietened again to hear me. 'I respect in my turn the wisdom of the noble Claudius,' I said. 'And, respecting his wisdom, this is how I answer him. And all of you.' I raised my voice till it rang under the long, dim roof. 'If a man takes up a sword, and goes with it to the market-place, and injures one of his fellows, then he will be punished; for that is right, and according to the Law. But if that same man wakes in the night, and hears thieves within his house, surely he will then arm himself to protect his property, and family. If he does not, will not his neighbours say, "See, he is either a coward or a fool"?' I heard the din rising again, and beat it down. 'Britannia is your house,' I said. 'And the thieves are at the gate. When, in all their past history, have Britons been either cowards or fools? I answer you myself. *Never. . . .* Let not such foolishness start now, in this place; and you yourselves its authors. . . .'

A rising gale of voices. But as many seemed to shout for me as against; and Tammonius, with excellent timing, put in the last word. 'I myself, as you have heard, still hold the north,' he said. 'Hnaufridus holds the south and east. Over the months to come, the Praefect Sergius Paullus will travel among you. Where he sees your walls are weak he will call on you to strengthen them. Where he sees your young men ill-prepared for war he will correct their faults. Where he sees your vigilance and energy have borne fruit he will send good word for you to the Emperor. Meanwhile we will pray; for the success of Stilicho, and the deliverance of Rome.' He inclined his head courteously towards the Bishop of Augusta. 'We will pray as citizens and as Christians; and we will arm ourselves to fight. Now let us have an end to this; for we are weary, and wish to rest. We will talk again this afternoon, discussing the other things that have brought us all together.'

In that way I landed myself with an arduous and thankless task.

The Council dragged on for another week. I didn't wait for the end; I took my leave of Tammonius and hurried south again. I was keen to implement a new idea. Along the coastline of Brigantia chains of semaphore towers had been established. I doubted whether Hnaufridus could spare the labour to copy them in the south, but there seemed no reason why strategically sited beacons couldn't be used to direct fast-moving bodies of troops. In that at least the Count seemed only too willing to co-operate. The parties detailed for the work set to enthusiastically, piling up heaps of faggots in what I would have thought were the least likely places. It was all very encouraging; though as I remarked to Hnaufridus, there seemed little point in embarking on too ambitious a programme of bonfire building till there was somebody, or something, to which to signal. My notion of highly mobile units had so far made little headway, and, of course, I was scarcely in a position to instruct the Count on the organisation of his defences. I spent a week or so at Portus Adurni, trying generally to familiarise myself with the area, before deciding to move north once more.

I travelled by easy stages to Camulodunum. The Celts followed uncomplainingly, content it seemed to amble about the Province indefinitely. By August I was in Lindum. My first duty call was on Claudius Felix. Despite his former opposition I found him vigorously engaged in strengthening and reorganising the town's defences. I stayed a fortnight, ostensibly overseeing the installation of catapults and ballistae. In fact the work was as new to me as most of the others. I could only try to recall and copy what I had seen in Gaul. The engines rested on platforms of brushwood and puddled clay, which serve to absorb the massive recoil shock. Their construction called for skills neither I nor the town aediles really possessed; we proceeded by trial and error, but the end results seemed fair. I wasn't basically too concerned for the accuracy of the city artillery; in my experience I'd always found a strong circuit of walls to be deterrent enough for all but the most determined raiding party. If seaborne, barbarians are usually more concerned for the safety of their ships; they tend to bypass fortifications in search of easier plunder. However, an effort was at least being made; I complimented Claudius warmly on the work—how far that moved him I couldn't say—and posted north again for Eburacum.

Duke Marcus welcomed me enthusiastically. The news from Gaul, he told me, was generally good. The present location of the British troops was uncertain, but the enemy were being contained on all major fronts. Stilicho was in Italia; I was more concerned with the whereabouts of Alaric and his Goths, but on that point no exact information was available. Sooner or later a confrontation seemed inevitable; from my point of view it couldn't happen quickly enough. Britannia's walls might be strong, but I'd seen at first hand how completely her emergency forces lacked cohesion. All that was needed was another concerted barbarian attack, such as had already taken place in the time of Maximus. We would be defeated in detail, and pushed into the sea.

I was particularly keen to see the Wall, still reckoned by many the greatest of Hadrian's achievements. The Duke rode with me himself on a short tour of the outposts of his command. Until then I hadn't appreciated the size of Brigantia. From Eburacum we rode two days north, for the most part across sweeping, desolate moorland, before finally coming in sight of the straggling township that runs almost without a break from one side of Britannia to the other. Beyond was the Wall itself. By nightfall we were at Segedenum; the following day we started west, along the huge fortification.

I rode in silence for the most part, absorbed by new impressions. Milecastle after milecastle was passed and still the great barrier climbed and soared ahead, clinging to the faces of precipitous slopes, climbing to cross crag after lonely crag. The smoke from many cooking fires threaded into the sky; high above, hawks hung motionless, mere dots against an immensity of blue. Beyond was the heather, purple-gold with the first glow of autumn; beyond again, shadowy and vague, the hills that ring Caledonia. The air was clean and rushing, sweet as wine.

Tammonius finally called me from my reverie with a remark about the defences. Resting each end on the sea, the Wall was easy enough to outflank; any determined force, well equipped with boats, could force a passage to the south. It had happened in the past times enough, and would certainly happen again. Some such thought had occurred to me; but, of course, the Wall had never been conceived as an impregnable barrier. Rather it had been a line of demarcation, splitting the perennially unstable Brigantes from their northern allies. The great ditch that had guarded it to the south was for the most part filled in now; causeways had been built across it at innumerable points, to give access to the Wall and Military Way. I threw a casual question about the current loyalty of the north Britons. Tammonius answered with a shrug.

As we moved west the Wall grew lower and broader, and signs of past destruction were more frequent; tower after tower had been roughly overthrown, even more crudely rebuilt. Here too I was surprised to see civilian traffic moving unconcernedly to the north. The Duke explained that the tribes immediately beyond the Wall were well disposed towards the Empire, having been bound by treaty alliances to keep the peace. An annual allowance of grain was made to them; their Kings wore purple, as befitted Roman generals, and were very proud. It seemed a precarious arrangement to me, but better, I supposed, than none.

At Luguvalium the Duke turned back. I pressed on south, for Deva. The weather held; mornings were misty, the days still and warm. We came in sight of the town, the old headquarters of the Twentieth, late one afternoon. The barracks, workshops and massive granaries still stood as they had been left, but Deva had held no troops now for many years.

The Celts disposed themselves in one of the disused barrack blocks, piling in five or six men to a cubicle, though God knew there was room enough to spare. I carried my gear to the centurion's quarters at the end; the empty, echoing Praetorium was altogether too much of a mockery. Towards evening, impelled by a restlessness I could no longer control, I made a tour of the place. I think it was then for the first time a real sense of desolation came over me. I climbed finally to the empty rampart walk, stood staring west towards the distant mountains of Siluria. Below me, closer at hand, was the harbour, still with a gaggle of shipping, the sandy estuary stretching away into the dusk; at my back were the long clusterings of roofs, the parade ground deserted and quiet. Across it the last of the light lay calm and golden. Beyond the nearest of the barracks smoke rose from where the Celts prepared their evening meal. Other blocks already housed families of squatters, wandering tinkers, peasants and the like; and a mule train, headed south, was resting for the night. As I watched a stray dog loped from somewhere, turned into the shadows of the huts and was gone. A door banged, raised a little clapping echo. A man passed, trudging below me; he didn't look up, and I found myself toying with the odd idea that I was invisible, a ghost returned from some sad Hell to the scene of former triumphs. For three hundred years the men of the Twentieth, and their families and wives, had lived here and had their being. Now they were gone. Momentarily it seemed I heard the jingle of harness, the rising tramp and clatter of feet; but it was only the wind.

There was a voice behind me, calling excitedly.

'Sir . . . sir . . .'

I looked up. It was Riconus, dusty and dishevelled, but beaming all over his face He came bounding up the last of the steps to me; and I stared in amazement at the thing he held out. Tarnished it was and uncared-for, but unmistakable; a great staff, bronze-tipped, with the insignium of a charging boar.

'Where the Hell,' I said slowly, 'did you find that?'

'In the old chapel,' he said. 'Slung in a corner, under some junk. There was an old man, real dodderer. Reckoned they marched without it. Said the Christos banner was good enough. . . .' He was polishing, vigorously, with the sleeve of his tunic. 'They left the Standards,' he said. 'Even left the Standards. Reckon they left their luck as well. Look at that; it'll come up a treat....' He glanced at me appealingly. 'Can we keep it, sir? We've never had one of our own....'

'Good God,' I said, 'what are you thinking of, Riconus? That isn't ours.'

His face dropped. He stood cradling the boar against his massive chest; and in spite of myself I had to laugh. 'I haven't seen the thing,' I said. 'If you want to collect baubles, that's your affair. But just don't wag it too often under my nose....' I walked towards the steps. 'I fancy some of that stew down yonder,' I said. 'Let's get to it, before those hungry wolves have cleared the lot.'

I glanced round me one last time before descending, but the circuit of massive walls had already slipped into the dark.

Chapter fifteen

I wintered at Deva, painfully aware of its insecurity. To the east were the hills that form the high spine of Brigantia; west and south lay the impenetrable fastness of the Silures. The fortress stood squarely in the gap, effectively defenceless, a standing invitation to seaborne attack; it was the weakest link I had yet seen in the defences of the Province. I wrote urgently to Tammonius, stressing the danger of the situation. I was rewarded by the appearance in November of a numerus of cavalry, detached from service on the Wall. They were followed by a ragtag and bobtail of archers, Syrian, half-caste and British, and one of the Duke's last regiments of German infantry. When they arrived I breathed a little easier. At least I had some solid troops, that I felt I could rely on. The archers and infantry I detailed for garrison duty; the cavalry rode with me on patrols of the surrounding country. They too were a mixed bag: Hispanians, Belgic Celts and a few Franks. I preferred to keep them under my eye.

Winter clamped down, howling and grey. With it came the raiding parties I had anticipated. None of them wasted time besieging Deva; one and all pushed inland, outflanking the one manned position. We burned their boats, regularly and methodically. Sometimes we brought the owners to a stand, but usually not. They simply vanished, into the surrounding wilderness. The country was too big, the defending forces too small. I sent messages to the inland towns instructing them to look to their walls, while an increasing stream of refugees began to pour into Deva; peasants for the most part, whose holdings and villages had been burned. There were few rich estates in the area, but the rank and file of the Britons suffered sorely.

The new arrivals brought fresh problems in administration. For the most part they were too dispirited to do more than huddle in the unheated barrack blocks, waiting for death either from privation or enemy attack. I ordered the Praetorium buildings prepared for them, the stoke-holes fired. They had to be herded to the new quarters with the flats of swords. After that I had to further subdivide my men into armed foraging parties. Grain stocks were low; we had barely enough for ourselves, there was nothing to spare to feed the extra mouths. Messages flew again, to Eburacum and the Wall, begging supplies, but there was little to be had. Only Luguvalium answered. One grain cart reached us, miraculously; the rest were waylaid, their teams and drivers senselessly butchered. The refugees began slowly but surely to starve. I had to close my ears to the endless sound of misery. There was nothing further I could do.

A little news came in from the west. Segontium, miles away fronting the Isle of Mona, was held by a strong garrison, but most of the tribesmen had retreated to the hills. From there they waged bitter, unrelenting war on their enemies. There were battles in the mountains that will never be retold, massacres that will never be mourned. While gale after gale roared in blurring the outlines of the land with a creeping blanket of snow.

The New Year brought at least one relief. The raiding stopped; nothing now could cross the Hivernian Sea. My sentries were blind; for days on end we couldn't see farther than a javelin-cast from the walls. The guards came down with their cloaks frozen into folds, their beards stiff with ice. I clung stubbornly to my duty rosters. In this land news travelled by means I couldn't even guess at. The enemy might be invisible, but my walls must be known to be manned.

About that time I had what was probably my closest brush with death. Riconus caught the assassin, an Hispanian, creeping into my quarters, a drawn sword in his hand. Whether he had been hired, or whether the cold and privation had worked too powerfully in him, I never knew; the Celt's methods, though effective, ruled out the possibility of interrogation. The head was displayed for some time above the gates as a warning to future malcontents, and I moved in with my men. I understood now their insistence on crowding together to sleep; one and all, they were masters of self-preservation.

But the hero of Deva was undoubtedly Valerius. I had been uncertain at first about taking him on, but through that grim winter he proved his worth time and again. Nothing seemed to daunt him for long and his energy was limitless. He headed foraging expeditions, drew up the rosters, re-established a camp hospital and actually chivvied a handful of the better-favoured refugees into forming a species of Town Guard. He grew quite proud of them, paying their grain rations from his own carefully-hoarded stores. In time the idea caught on; the 'Tribune's Patrol' mounted the walls alongside the Germans, armed with clubs and staves. How they would have reacted at sight of an enemy is hard to imagine; perhaps it was as well they were never put to the test.

It was Valerius too who rode to Manucium to bring back a sorely needed British farrier; while in between his various self-allotted tasks I found he had set himself to learn German. Though he disapproved in principle of aliens on British soil—he confided as much to me, one night at supper—he had decided the Franks and Burgundians under my control were first-rate soldiers. 'And after all, sir,' he said, 'an officer ought to be able to deal directly with his men. That's obvious. . . .'

March brought a lessening of the strain. We saw the sun, it seemed for the first time in weeks; and I mounted a series of well-armed patrols, determined to take the offensive against any parties of raiders still in the vicinity. No contacts were reported; it seemed the land round about was clear. A despatch from Tammonius confirmed that the northern sector was also quiet. I risked a trip to the south, making the long ride to Portus Adurni. The news that had filtered up from the Saxon Command had been universally bad; it seemed Hnaufridus had had his hands more than full.

I found him in a high rage, stamping round his quarters like an infuriated bull. 'Ah, you come,' was his greeting to me. 'I wait for you. Tell these'—the phrase he used escaped me, but was evidently far from complimentary—'I no longer keep their country. Tell them I go back to my own land, over the sea.'

His complaints, narrowed down, centred on the generally unco-operative attitude of the southern British. One and all, the inland towns seemed to feel the levying of emergency forces was no concern of theirs. They sat plumply, waiting for protection; or, as the Count put it, like chickens on a spit, ready for the roast. His depleted garrisons, already over-extended, had been no match for the fleets of pirates that since my last visit had infested the coast. The military forts, solidly built and heavily armed, had for the most part weathered the storm, but inland it had been a different story. Venta, only a handful of miles from the Duke's own base, had been insolently besieged, and Noviomagus severely battered. The forces withdrawn to cope with the attackers had left wide gaps in the defences. The original trouble had been caused by Saxon war parties, but Scoti had instantly appeared from the west and north, and confusion had turned to disaster. Lemanis had been reduced and its garrison—among them some of the Duke's best troops—massacred, while villages and farms had been looted and burned over a wide area. It looked as if what I had feared was beginning to happen; the barbarians were once more acting in concert. Units of the British Fleet, hastily despatched from Gaul, had claimed moderate success, but by then the damage had been done. An entire fleet of German and Scotic war boats, all heavy with plunder, had been sunk off Dubris without trace; that added if anything to Hnaufridus' grief.

I wrote a despatch for Tammonius, reporting the situation and begging the release of troops from the north to stiffen the sector and make good its losses. I didn't expect to be answered, and wasn't. I spent the rest of the season riding, from Venta and Portus Adurni as far afield as Camulodunum. A dozen times the scene in the Great Basilica repeated itself almost word for word. Maybe there were results. They were too few, and all too sporadic, for me.

I was conscious that my attitude was hardening. At first, remembering the network of spies that exists in every Province, I had taken every opportunity of stressing the virtues of Honorius and his concern for all things British. I think Valerius had come to know the speech by heart. Now I no longer bothered with it. The theme of my addresses was simple and blunt: 'Arm yourselves, or die.' I'd likewise ceased to worry too much over the precise limits of my authority. I had a more accurate idea of the layout of the Province now, gained by the endless travelling. A vast triangle of territory, from Augusta north to Deva and west to Glevum and the Sabrina, lay open and virtually defenceless. Within it were probably a few thousand limitanei, a handful of Regular detachments, and myself. Whether I liked it or not, I was one of the commanders of the Province.

Things came to a head at Caesaromagus, on the coast road from Augusta. Outside the town I found an Imperial posting station garrisoned by a dozen scruffy cavalrymen. As Valerius had warned me, such detached auxiliaries were usually more concerned with raising wheat than the performance of their duties. Looking back, I don't suppose they were wholly to blame; most of them hadn't seen a penny of their pay for years. But I had neither time nor inclination for niceties of moral judgement. When I paraded them nobody could account for his full equipment, while three of the creatures owned neither horses nor weapons. The animals had died, they told me, with nicely barbed insolence; while their swords and armour had rusted clean away. It was the climate.

I surprised myself, and them, by clapping them in irons. When the anger passed I wrote to Hnaufridus for a directive; after all the men were, in theory at least, under his command. His office took their time about answering. I kicked my heels for a week—I could have been better employed in a score of places—before receiving the laconic suggestion that beheading seemed as good a solution as any, but it was really up to me.

I'd never had to answer for men's lives in cold blood. I didn't relish the prospect now. The prisoners were pitiful-looking villains, unshaven and unkempt; I doubted they even understood the severity of their crime. But I was fairly caught. Some sort of example had to be made or the word would spread, quicker than I could ride, that Roman authority was dead. In the end I commuted the sentences to flogging, and paraded the rest to see the punishment carried out.

Wounding in battle is horrible enough, but a flogging is something different again. The first strokes cut deep, and the blood starts to stream; then the scourge begins falling on flesh already torn. At first the victims scream like animals. Later they are dumb. There's no sound but the thudding of the whips.

I stood to one side, arms folded, Valerius a pace or so behind me. I would have called out to stop the thing. I found I couldn't. My throat constricted; the ground round me began to flicker and heave. I turned to Valerius. His face was a dirty white, glazed with sweat. He was weaving his head, it seemed unconsciously; swaying, and drawing himself erect.

It was too much. One couldn't have senior officers collapsing in too broad a swath. I barked, 'Fall out, Tribune. . . .' The sound of my voice broke what had seemed like a horrible spell. 'That's enough,' I said. 'That will do.' The prisoners were released from the frames to which they had been lashed. They lay shaking, red things that might perhaps live. I dismissed the parade, rode the rest of the day in silence. I was remembering the gesture Stilicho had prevented me from making. I had been on the point of riding to Mediolanum, to deliver myself up. To that. I'd thought the decision mature and noble. Now I realised it hadn't been a decision at all. One can't make decisions until one knows the relevant facts. In Massilia I hadn't seen a flogging. Now I had.

I headed for Camulodunum. The Senate there had passed a resolution condemning Roman protection as ineffective and had let it be known they were considering overtures to the Saxons themselves, hoping that in return for land grants the pirates might undertake the town's defence. I hope the address I gave them is remembered. It began with a description of the death of the Red Eagle. It's as good an illustration as any of the barbarian sense of fun. I ordered the recruitment of five hundred men to serve as a garrison with the status of limitanei, giving the curiales a month to create an effective force. The unit was somehow enrolled. It lacked armour and most of the men had no swords. I requested the immediate despatch of a sum of money to Eburacum and wrote to Tammonius asking that his armourers, and any more who could be pressed or cajoled into service, set to at once to make the deficiency good. I left Valerius to oversee the training of the new force and rode west again by way of Verulamium. More trouble had flared behind my back, in Britannia Prima.

At least there was no fault to find with the defences of Verulam. It was by far the smartest, best-cared-for city I had yet seen; and its massive walls and gates were heavily manned. I stayed overnight at the home of a curiale, Pacatianus. There was, he told me, no lack of money in the Colonia; in fact an extensive building programme was currently in progress. I would have stayed longer had I been able, savouring the unexpected air of confidence and prosperity, but there was no time. The situation in the west demanded intervention; I rode by long stages to Corinium.

The problem was with the laeti, Germanic farmers who had been established round Glevum and Corinium and along the banks of the Sabrina. In theory such settlements acted as buffer zones in the absence of regular troops. A similar scheme was operating in the north. There it seemed to be functioning successfully, but in Prima it had already bred a state bordering on civil war. A series of ugly incidents had taken place; crops had been burned, a man or two killed, and tempers were running high on both sides. I arrived myself in a scarcely simulated rage. The Province had troubles enough without such pointless bickering.

For sprawling, ill-kept immensity, Corinium was second only to Augusta. I took up my quarters in the town and requested, through the office of the Praeses, a full meeting of the Curia. It was called for the following week; and I sent messengers to the scattered German settlements instructing them to send delegates to state their case. In the event, none arrived; I was forced to speak to the Curia in their absence. I had spent the intervening time making enquiries of my own. It seemed most of the trouble had been started by the British. There was widespread resentment at the requisitioning of land for foreigners, while for their part the laeti were obviously prepared to defend their newly founded homesteads with their lives. I told the curiales, sharply enough, that the blame lay at their own door; I stressed the value of laeti in the overall scheme of defence, and ended with a dire though unenforceable threat to garrison the place with Burgundians. That at least was effective. A motion was passed pledging me the support of Corinium. I suggested in my turn that the town's resources be employed in the creation of a force of limitanei similar to the one I had established in the east. If the local young bloods needed to work off excess energy they would be better employed slashing at practice stakes than barbarian settlers, whose only fault was probably their lack of Latin.

It was harder to win the confidence of the laeti. I managed finally to convene a tribal council on neutral ground a few miles from Corinium. I attended in full-dress uniform, absurd though I felt the whole thing to be, with half a dozen burly Celts as bodyguard. At least it made a colourful display; the Germans, big, slow-moving men for the most part, seemed suitably impressed. They were diffident and awkward at first;

eventually they realised they had a genuine chance to air their grievances. The speeches went on for a day and a half before I pronounced judgement. 'This armband I wear,' I said—the golden torque had once more worked its charm—'was given me by a chieftain of your own people, a great prince in his land. So you know that no stranger speaks to you, but one very like yourselves. By it I swear that your interests shall be my interests, your wrongs my wrongs. Harm done to you will be as harm to me. In return you must promise to keep the Law, and not take up arms against your neighbours. For the crops that have been destroyed, you will be given seed corn in the time of planting. I cannot restore your lost ones to you, but their blood money will be paid in gold. By this means their ghosts will be placated, and will not trouble you at night. May the Vanir guard your fields,' I finished haltingly, 'and the Aesir look on you with kindness.'

I was reprimanded, of course, by the office of the Praeses for my high-handedness; but as I pointed out, at least the appearance of peace had been restored. The reparations were duly made; it seemed the threat of Burgundians carried some weight, even with a Roman governor.

Summer was once more drawing to an end. Mornings and evenings were misty, the air filled with a faint pervasive tang like the burning of dead leaves. I would have liked to rest, but I still seemed to be working against time. To the west, ramshackle and undefended, lay Glevum, fronting the Sabrina and the mountains of the Silures. I spent some time in the town, trying to bring its inhabitants to an understanding of their danger. I made little headway. The quinquennales were away, I was informed; what remained of the Town Senate was hopelessly apathetic. Regular troops seemed the only answer. There were none to be had. I considered the possibility of bringing Germans down from Deva. In the end I dismissed the idea. Neither the Duke nor Hnaufridus could be expected to tolerate any further drain on their garrisons. We had reached our limit.

Once more, circumstances rendered me helpless. Round me lay a rich countryside. Sheep grazed the downs and rolling hills; wheatfields lay ripe and golden in the sun, villas nestled against the sheltered slopes. To all this, Glevum was the gateway. But Glevum wouldn't fight.

I retired to Corinium. At least its walls were in good repair. I saw to the raising and training of the garrison. Day after day the place rang with the clash and rattle of practice weapons, the measured tramp of feet. It looked, and sounded, impressive. I wrote to Tammonius again, begging javelins and swords. If he couldn't send me troops, at least he could open his armouries. A few weapons and breastplates finally arrived; barely enough to equip one man in three, but better than nothing. Shortly afterwards I held a parade of the new-fledged soldiery. The curiales of the town were delighted; but the occasion served, if anything, to increase my gloom. I had already seen how the Scoti fought. It would take more to stop them than pennants and gaudy cloaks. I dictated a full report on my activities, marking it for the personal attention of Stilicho, and detailed a man to ride with it to Augusta on the first stage of its long journey. Maybe it arrived, maybe not. Either way it made no difference to Britannia. I was asking Mediolanum for a Legion. It was a joke in bad taste.

Valerius rejoined me in December, armed with a massive screed covering every aspect of his work in Camulodunum. He told me the Saxon Shore was for the moment quiet. I wondered, wearily, from which direction the next inundation would come.

For the present there was nothing more I could do. In not much over a year I had ridden almost the entire circuit of the Province; and I was deadly tired. I decided to winter at Corinium. At least the town still possessed a few civilised amenities. A letter from Isca of the Dumnonii, requesting an authority to raise a regiment of its own, I passed to the office of Hnaufridus. Maybe with the spring I would travel into the far south-west, but not before.

The first snow fell, muffling the rutted streets, mantling the leaning, steep-gabled houses with white. Christmas came and passed; I shivered in church with the rest, celebrating the Birth of the Lord. The New Year saw us in the field again. Raiding parties poured in from the coast; every day brought some fresh news of disaster, while the trickle of refugees flowing into Corinium thickened to a stream. It was Deva all over again; except that we were numerically too weak to mount anything like an offensive. We sallied cautiously, keeping the walls at our backs, to engage such of the pirates as came within reach; but our efforts were little more than a gesture. The enemy learned to give the garrison a wide berth; elsewhere, along the coast and southward to Aquae Sulis, they burned and harried with impunity.

Spring once more brought a lessening of activity. Rumour had it the paramount King of Hivernia was preparing a massive onslaught, and had withdrawn his men. I had no means of checking the report, but at least the lull was a welcome breathing space.

The new season also brought an unexpected letter. I opened it curiously when it was sent in to me. It was couched in stiffly formal terms; it invited me to Censorina, a villa some ten miles north-east of the town, to a dinner to be given in honour of C. Sergius Paullus, Palatine Praefect and Commander of the West.

I didn't care much for the tone of the address. They would be naming me Count of Britain next. Circumstances might have forced me to make decisions that properly belonged to a higher authority, but I had no interest in claims that would excite the immediate unwelcome curiosity of the Emperor. I made discreet enquiries, through the offices of the ever-resourceful Valerius. Censorinus, the owner of the property, was apparently one of the wealthiest Britons in Prima, with business interests as far afield as Belgica and southern Gaul. What he could want with me I had no idea, but all considered it seemed best to attend. I delivered my acceptance, and set out some days later to ride the few miles to the house.

Valerius attended me. The sun shone from a deep, clear sky; the trees of the many coppices were hazed with fresh green, but the wind was still sharp-edged. It was midday before we discarded our cloaks; shortly afterwards we came in sight of our destination, and both reined.

As I have said, the area round Corinium was dotted with wealthy villas, but Censorina was the biggest I had seen. From the trackway we were following the land sloped up gently to a curved and wooded ridge. In the tongue-shaped valley thus formed lay a great rectangular building. The nearest wall was pierced by an imposing entrance arch of stone; through it I glimpsed lawns and gravelled walks. Over the walls rose neat lines of red-tiled roofs; beyond again, crowning an eminence a few feet above the rest of the complex, were the columns and pediment of a graceful little temple. The whole place breathed an air of order, prosperity and peace.

Valerius stared at me and shrugged. I touched heels to my horse; a few minutes later we rode, harness jingling, beneath the gateway, dismounted in the wide courtyard beyond. We were met by a slave in a tunic of crisp white linen. He greeted us courteously in good Latin, welcoming us on behalf of his master. The Domina, he said, had ordered the bath-houses prepared; everything was ready, if we would be good enough to follow him.

A colonnaded walk stretched round three sides of the place. I followed the guide, nodding absently at his murmured comments on the weather and the state of the roads. Wealthy, Censorinus obviously was; his establishment reflected a level of taste and elegance I hadn't so far seen in Britannia. We were offered, to my surprise, a choice of refreshment; for in addition to the normal suite the house possessed a Spartan plunge. Valerius opted enthusiastically to try the process; I settled for the more sedate arrangements to which I was accustomed.

The bath-house was neat and clean, its floors decorated with mosaics of doves and leaves. I was divested of my weapons and clothes, spent an hour stewing in the hotroom before submitting myself to the attentions of a masseur. It was curious to sit chin-deep in scalding water and hear birds singing in the nearby woods. It seemed already the place was throwing a spell on me, a pervasive charm of peace.

My tunic, as I had expected, was returned pressed and warmed. Valerius, pink-cheeked and healthy-looking after his exertions, was waiting for me in an anteroom. A slave appeared, bowing, to conduct us to the presence of our host. We followed him on along the colonnade. Early-evening sunlight lay richly on the lawns. Doves called from the trees; from closer at hand came the feral yell of a peacock.

The triclinium in which we found ourselves was in keeping with the rest of the house: a wide, low room, its walls decorated with elegantly painted plaster. A mosaic, with figures of the Seasons executed in the same cheerful Provincial style, occupied most of the floor; beyond, couches and chairs were grouped round a massive inlaid table. White-robed slaves stood about unobtrusively; a number of lamps was already burning, their warm glow competing with the last of the sunlight from the door.

The company, small as it was, was already seated. Censorinus rose to greet us. I saw a man of above-average height, but so slimly and delicately made that at first his tallness wasn't apparent. Thinning sandy hair combed flat across his scalp accentuated rather than concealed the long line of his skull. His face was equally smooth, the skin drawn tight across the cheekbones. His eyes, cold and indefinite-coloured, never seemed fully to meet mine; his voice was cultured and flat, with a trace of sibilance. 'Allow me to present,' he said, 'a business associate of mine, the Senator Gnaeus Gratianus; and my wife, the Domina Crearwy. The Palatine Praefect Caius Sergius Paullus.' Then, directly to me, 'You are most welcome, Praefect. I've been looking forward to our meeting. I hope our people have been looking after you satisfactorily?'

I murmured something appropriate. My eyes were on the mistress of the house.

She sat, correctly, on a carved, high-backed chair. Beside her was a slave, a piquant, wild-faced little thing with long dark hair reaching to her waist. The Domina herself was tall, I guessed; as tall as or taller than her husband. She wore a light blue robe, gathered at the shoulder by a complex golden clasp. Her hair, almost equally long, glowed softly, the colour of rich corn. She wore it loose, restrained by a simple fillet. Her face, neither young nor old, was long in jaw and nose like the muzzle of a cat, yet delicate, with a humour that quirked the corners of the firm mouth, and eyes as blue and vivid as the dress. They met mine and locked, moved away and returned; and it seemed for a confusing moment I was back in the house of Paeonius, seeing the heads of Gods loom from the dark. Calm heads, wide-eyed and inscrutable; heads of Pharaohs and their Queens, women who were cats and tigers that were men. In each a Line, discovered over and over in the stone, that is immortal.

Beside me Valerius had already disposed himself on the nearest of the couches. I sat, clumsily. I turned my attention briefly to the other guest, a balding, heavy-jowled man in a tunic of scarlet. He was silent. He was to remain silent throughout the meal; as far as I can remember, he never spoke a word. But his eyes, liquid and dark as the eyes of Censorinus were pale, watched incessantly.

The master of the house clapped softly. Dishes of aperitifs were produced. I hadn't seen food like it since leaving Gaul.

There were oysters with cumin sauce, bonito, truffles, mushrooms and fresh eggs; a tree fungus with a delicate, woody flavour, platters of plump stuffed dormice. I was being shown, it seemed, just how a wealthy Briton lived.

Better was to follow. The bowls of honeyed wine were removed, the heavier vintages of First Tables took their place. With them came peacock rissoles, a sucking pig garnished with early vegetables, and a spectacular delicacy I at least could have done without; wombs of sterile sows, served grandly with silphium, liquamen and vinegar. Beside me Valerius ate steadily, reduced at last to silence; while through it all Censorinus droned on, voice rising and falling in a monotonous rhythm. We discussed the state of local farming, the new troop dispositions in the Province, the Empire at large; and succeeded in saying, as far as I can recall, nothing at all.

I had been wondering about her name. What was it, Crearwy? It had a strange ring. Barbaric. Something else too was pricking at my consciousness. A sense of non-surprise, almost fulfilment. It seemed I already knew her face; the forehead, narrowing slightly from the cheekbones, the level, unplucked brows. Which was curious, for I had certainly never met this woman before. Maybe, though, she had once watched up from a mosaic; while her cloak blew, baring the broad, lovely hips, flowing in a non-existent wind. And a fountain splashed and sang.

'I'm sorry,' I said. 'Your pardon?' Censorinus had been talking again; and I hadn't heard.

'No, sir,' I said. 'I'm not from Latium. My home Province was Hispania.'

'Was, Praefect?'

It was a surprise to hear her voice. As if stone could indeed speak, a little huskily.

'Yes, Domina,' I said. 'It's been many years since I left.'

'Then we're both strangers to Britannia,' she said. 'For I am Scotic. My father owns lands in Dalriada, beyond the Wall.' She inclined her head. 'My husband is from Brigantia; it's been a long time since either of us saw our homes.'

There were many things I could have asked her, but Censorinus gave me no chance. His voice flowed on remorselessly, still with its smooth cadence. He was talking about the Province again, probing and questioning delicately. Tammonius Vitalis he knew well, had known for many years; the Count of the Saxon Shore he hadn't had the pleasure of meeting. What sort of man was he? An efficient soldier, by all reports. What loyalty did he command among his garrisons? And what was the state of the troops now holding the Wall? Could they withstand a concerted attack? He had heard a little of my activities in Deva. At one time the garrison had been deserted; he took it this was no longer the case?

The food was superb, the wine more than adequate; I was rested, refreshed and at ease. Under the circumstances quite why I became so unconscionably annoyed is hard to say. Maybe it was his manner, the glacial calmness, the bland authority with which he directed the talk; certainly I hadn't come here to be quizzed like a new Tribune giving his first report. I found myself answering more and more shortly, drinking deeper from the endlessly replenished wine-cup at my elbow. Trays of sweetmeats were passed; stuffed dates and wine cakes, figs and fruit. I was aware of Valerius watching me narrowly; aware of the startling woman at the end of the table; aware, too, of the mosaic of the Anadyomene. Only here were no fountains. The little alcove where we sat was becoming stuffy with heat.

Night had fallen during the long meal. More lamps were brought in, and still the rigmarole went on. Crearwy, it seemed, had finished with eating; she sat expressionlessly, watching down at her plate. I spat a pip deliberately into my palm, and made an effort to turn the conversation. I had heard, I said, that my host had business connections in Gaul. I had served in the Province myself for some years, before being ordered to Britannia.

Yes, this was true. The Domina affirmed it; Censorinus seemed undisposed to comment. Her husband's interests were varied, but mostly they lay in mining. Had I perhaps had some experience . . . ?

I was drinking at the time. I coughed, feeling the wretched stuff sting the back of my nose. I set the cup down carefully.

'Yes,' I said. 'I have had some experience of mines.'

Then I might be able to advise Censorinus on several technical matters that had been causing concern. But for the moment . . . the talk veered back, inexorably, to those subjects I was keenest to avoid. What, for instance, were the Praefect's reactions to this new notion that seemed to be gathering ground, this idea of an expeditionary force to Gaul? A British force, British led. To all intents and purposes, the Province had been deserted; but at all costs, surely her links with Rome must be maintained?

'I know nothing of such a force,' I snapped. 'And nothing of such a scheme. Nor, I warrant, does Mediolanum. If these rumours are circulating, and personally I haven't heard them, they're baseless. Furthermore, they don't even make sense.'

He had stopped, chin raised and blinking, affecting an air of mild concern. 'But surely, Praefect,' he said, 'is it not an obvious move? In face of the threat from the Rhine?'

The wine seemed to catch fire inside my head. I slammed the cup down on the table. 'And I say it's lunacy,' I said. 'And dangerous lunacy at that. Ask your friend Tammonius, if you don't want to take my word.' I ticked off points on my fingers. 'In the time of Severus,' I said, 'Albinus stripped this Province of troops to fight in Gaul for the throne. What was the result? The Wall was overrun, the country wasted as far south as Eburacum, while Clodius Albinus died. In Diocletian's time, Carausius and Allectus usurped the Purple. The result? The Wall was overrun, the country laid waste again. Carausius and Allectus died. In Theodosius' reign, your own time, for God's sake, Magnus Maximus declared himself Augustus. The Wall was overrun, the Province laid waste; and Magnus Maximus died. You know the results, you saw them with your own eyes; you all still smart from the Chain. What do you want, another little King? Can't Britannia ever learn a simple lesson? How often does a plain fact have to be repeated before it's driven into her skull? *She can't hold Gaul. . . .*'

If he was nettled he didn't show it. 'She has held Gaul,' he said. 'And Hispania.'

'She didn't hold Hispania. She was tolerated in Hispania.'

Crearwy said anxiously, 'Surely the trading routes must stay open. The Province has to trade to live.'

'No, madam,' I said. 'The Province has to trade so that her tradespeople can live.' I stared round the table. 'How long do you think you'd be safe yourselves if the Wall went down again? What have your children done, to have a Fate dragged on to their heads?'

Censorinus said, 'Britannia is stronger than she looks.'

'For the past twelve months,' I said angrily, 'I've done little else except ride her borders. I know her strength.'

He said mildly, 'She can raise more men.'

'Where from? Brigantia?'

For the first time I thought I saw a glimmer in his eyes. If it was there it was instantly veiled. 'Yes,' he said, 'at a need. The North has been effective in the past.'

I wasn't going to have his tribal pride rammed down my throat on top of all the rest. The words were out before I could stop them. 'Effective at stabbing a Legion in the back,' I said. 'I certainly agree.'

He stared at me palely. He said, 'We were being oppressed.'

'You were being protected.'

'Protected? From what?'

'Among other things,' I said sweetly, 'the effects of royal adultery.'

There was a silence. Then he rose. Valerius swung his feet from the couch, looking alarmed, but Censorinus merely spread his hands, benignly. Our discussion, he said, must be postponed, interesting though it had proved. He hoped to resume it at some future date. For the moment, several pressing matters demanded his attention. If I would excuse him, he had arranged a little entertainment; he hoped it would meet with my approval. He left the chamber, quietly, attended by a slave with a lamp. Gratianus stood in his turn, bowed silently and padded after him. The door swung to with a click. I sat back, still fuming, and met the Domina's eyes. They were unreadable, but her expression was nearly one of pain.

My cup was refilled. The door opened again. Musicians filed in, struck up a lively tune; and they came.

There were three of them. The Fates surely, or Harpies from old legend, raised by necromancy and chained to earth. One was old, one gap-toothed; all were fat. They bowed to the table and began to dance. Their feet slapped and whispered; breasts and bellies jiggled, gleaming through diaphanous robes. Dancing women I had seen, but none like these. The gesture, I supposed, was fitting; from a British aristocrat, to a ruffian from Rome.

'Valerius . . .!'

My bellow brought the capering to a halt. I was on my feet, and the wine was spilled, and the room was tilting and revolving. He ran for my sword and cloak. I buckled the weapon, saluted the Domina and left, heels ringing on the pavement. The horses were waiting, held by slaves with torches. I clattered through the archway, left riding like the wind. I didn't slacken till the great house was out of sight, hidden by the folding of the hills.

Chapter sixteen

Morning sunlight lay across the villa lawns. A gardener was working on one of the flower beds; the scrape and click of his spade sounded clearly in the bright air. Hens cackled from the stable block; there was the clank of a pail, the stamp and snort of a horse. A cart lurched out through the gateway, headed away down the sloping lane beyond. Crearwy bit thoughtfully at a stilus and watched it go.

She had had a chair carried out for her, and a little folding table. She sat now in a new white summer gown, a wax tablet propped on her knee. She glanced down, frowned, made an erasure and began again.

From the Domina Crearwy, she wrote, *to the Praefect Sergius Paullus, greetings.*

My husband has been called to Gaul, with the Senator G. Gratianus, on urgent matters of business. Before he left I begged him to resolve, if possible, the emnity between us. In this he was more than agreeable. I add my voice to his; I urge you to accept my apology, and the apologies of the household, for a slight which, however sharply felt, was unintended.

If your many duties permit, I would be honoured if you would once more join me for dinner, when I can perhaps make recompense, however small, for the injury suffered at our hands. Give your agreement, I pray you, to the bearer of this note. Any day and time you stipulate will be suitable to me.

She frowned, poised the stilus, hesitated then wrote rapidly again.

I shall be disappointed if you refuse, but will try to understand. Meanwhile I wait in the hope of our continued friendship. I'm sure there are many topics, unconnected with either politics or the State, that we would find mutually agreeable.

The Gods preserve you, and send your answer quickly.

I put the thing down and stared at it. Then I picked it up again. It was unexpected, certainly; also vaguely disturbing. More indiscretions have been committed to wax than ever attained the dignity of paper and ink. In the end I shrugged, and sent for Valerius.

'What,' I asked him, 'do you make of that?'

He studied the letter, frowning. Then he grinned. 'I'd say under the circumstances it was rather a nice gesture, sir,' he said.

'Under what circumstances?'

He looked troubled. 'Well, I suppose . . . the general set-up, sir.'

'Meaning?'

He compressed his lips. 'Nothing, sir. Just a figure of speech.'

'You're being singularly negative, Valerius. What would you do about it in my place?'

'In your place, sir?'

I said testily, 'Oh, for Christ's sake . . .'

'I'd go myself,' he said. 'Like a shot. Particularly if they're going to dish up grub like that.'

'Tribune, I don't really credit you with thinking with your stomach. You think I was wrong, don't you?'

He hesitated, saw me staring at him, and shrugged. 'I think you may have been a bit . . . hasty, sir,' he said.

'To be frank,' I said, 'I was as drunk as a newt. And that oily little Britisher succeeded in thoroughly rubbing me up the wrong way.'

He didn't answer, just sat looking carefully vacant. I hadn't seen fit to tell him I was suffering from a prejudice dinned into me in childhood by Marcus; and he knew nothing of the mines.

I got up abruptly. 'Bugger you, Valerius,' I said. 'All right, I'll make my peace. I'll go tomorrow. No, in two days' time. Satisfied?'

'Yes, sir,' he said. 'Will you be wanting me again, or not?'

'No, thanks,' I said tartly. 'If I'm going to be made to look an ass twice running I'd rather there were no witnesses. Would you ask Petronius if he can find that messenger?'

After he'd gone I picked the tablet up again. I sat absently, rubbing the smooth wooden cover with my thumb. Work was piling up and I'd been planning to travel to Isca later in the week. Now I'd have to put the trip off. I shrugged. I had to admit my conscience had been pricking me. It had been a thoroughly childish display; I still couldn't think what had got into me. I reached for a pen. The sooner the thing was resolved the better; I could get on with the job I'd been sent to Britannia to do. I finished the letter; then I sat and swore. Asking Valerius for his comments had been as pointless as it had been weak. From the first there had been no doubt about what I would do. I wanted to see the Lady Crearwy again; in that at least I was honest with myself.

The weather was perfect again when I rode for Censorina. Puffy white clouds chased each other across the sky; the air moved gently, smelling of summer. I took my time over the journey. This was good hunting country. I saw hares and deer; once a family of wild boars broke cover, trotted briskly across a distant skyline. The villa when I reached it lay bland and lovely in the sunlight. I turned my horse into the lane that led up to it. Twenty or thirty yards from the gateway a clump of bushes grew beside the path. As I was passing a voice rang sharply. 'Halt,' it shouted. 'Who goes there? Don't move, or you'll get an arrow through you.'

I reined, looking round for the aggressor. A rustling, and a small girl rose into sight, pointing a drawn bow in my general direction. She was I supposed nine or ten years old, a slim, brown-skinned creature with a tangle of lustrous dark-blonde hair. She wore a diminutive white tunic; her face was twisted into a determined scowl. 'Don't move,' she said again. 'Keep quite still, or I'll shoot.'

'You won't shoot anybody with that unless you learn to hold it properly,' I growled. 'Not even me. Your arm's too tense to start with; you should keep it slightly flexed, and only straighten when you loose. And for God's sake, girl, nock your shaft on the proper side of the stave. Here, come down and let me show you.'

A chuckle, and the bushes disclosed a second child, maybe a head shorter than the first. 'I told you you were doing it wrong,' she said. 'But you never listen.'

They thumped on to the path. 'Let me look at that thing,' I said. 'One of you hold the horse.'

The bow was passed to me. It was crudely constructed; just a greenwood stick, lashed optimistically at the ends to prevent it splitting. I flexed it and shook my head. 'This won't do at all,' I said. 'Where's your arrow?'

They handed it over. The flights were glued on tolerably well; the head was formed from a flattened nail. 'We made it ourselves,' said the elder sprite. 'Classius said he'd make one but he wanted paying. We hadn't any money.'

'Who's Classius?'

'The blacksmith. Are you an enemy?'

'I hope not,' I said. 'Do I look like one?'

'He's the soldier Mummy wrote to,' said the younger girl. She beamed at me. 'I heard Felix say about it. He's the doorkeeper. He always knows everything.'

I found a coin. 'Here,' I said. 'Give Classius this. Tell him it's worth at least three arrows. If you're going to ambush people you want your shafts to fly true.'

She said, 'Hello, Praefect.'

She had come down from the gateway. She was wearing a white dress, gathered by a girdle. The material hung softly, outlining her hips.

'I'm glad you could come,' she said simply. Then to the children, 'Nessa, Melinda, what do you mean by this? How many times do you have to be told not annoy our guests?'

I said, 'They weren't annoying me.'

Nessa, the elder, brushed the mane of hair back from her eyes. 'He gave us some money for Classius,' she said. 'We're going to have proper arrows.' They ran for the gateway, whooping.

Their mother flared at them.

'*Come here!* You'll get my slipper behind you in a minute,' she said. 'Where are your manners?'

They looked contrite. They said, 'Thank you, sir. . . .' Melinda grinned round a gap in her teeth. 'Will you teach us proper shooting, sir? We'll take you hare-spearing if you do.'

I said, 'I'll see.'

Once more the lamps glowed in their niches. But now there were no slaves. Only the dark girl I had seen before. Two couches were drawn up to the table. I lay on one; Crearwy reclined on the other. 'This,' she said, 'is another of my shocking habits. Do you think it's very bad?'

I said, 'Not if it's what the Domina wishes to do.'

She said, 'Are you always so terribly correct?' She paused. 'You're still angry with me, aren't you?'

'No,' I said. 'Not with you.'

She shook her head. 'He didn't mean it,' she said. 'Honestly. Not like that. He doesn't bear ill will.'

I held my peace.

'Did you like the meal?'

'Very much,' I said. 'Except the snails in milk.'

She looked disappointed. 'I'm sorry,' she said. 'I did it all myself. I thought you'd like them. I thought all Romans did.'

'I'm not a Roman. And I seem to be pursued by shellfish.'

'How do you mean?'

'My mother used to do them. The same way, in milk. Father swore they were good for my brain. It always seemed unfair. They swell so much they can't get back into their shells.'

She smiled. 'Tell me about your mother.'

'There isn't much to tell. She died when I was young.'

'What did she look like?'

I said, 'She was dark.'

'Was she an Hispanian?'

'No.'

She said, 'You don't like talking about your home much, do you?'

'Not really.'

'I'm sorry.'

A silence. Then she said, 'Would you like to hear Pelgea sing?'

'Does she sing well?'

She said, 'I think so.'

She turned and spoke a few words to the girl. The slave nodded, picked up an instrument like a complicated lyre. She struck a chord, gently. The sound was like rippling water, or the wind in the trees. Her voice was like her face, wild and sweet. I said, 'Where does she come from?'

'The north. She's been with me since she was tiny. She's a Pict.'

'I thought the Picts coloured their faces blue.'

'The Romans think all barbarians colour their faces blue.'

The song ended. I sat quiet. I said, 'Will she sing again?'

'If you want her to.'

The new song was longer; plaintive, and haunting. The

melody was strange. There was a lilt and shift to it like the movement of the sea. I said, 'What's it about?'

She pursed her lips. 'A girl, and a young man,' she said. 'It's a love song. The girl has died. Now her soul waits on the shore. It's dark, her eyes shine like moons. Soon her lover will come to her. He'll follow, across the sea.'

I said, 'She's waiting for the Boat.'

Her eyes widened. '*What?*'

I said, 'The Boat. From the Land of the Blessed. Tir-nan-Og.' She would have spoken again, but I held up my hand.

The song ended. The singer bowed, slipped back into the dark. The villa was very quiet. I drank wine, brooding. It seemed something was close to me, closer than in years. A Shadow, that still sought the West. Crearwy said, 'Who taught you about the Land?'

'My mother.'

'What was her name?'

I said, 'Calgaca.'

'Oh . . .'

It seemed I could talk to this woman, and be at ease. That in itself was odd. I said, 'We were very close. She was very . . . gifted.'

'In what way?'

'In many ways. She knew the Future.'

For a moment I thought she shuddered. She said, 'She had the Sight.'

'I don't know what she called it. It was a strange thing. My father would never discuss it. Do you have it?'

She said, 'No, thank the Gods.'

'Why?'

'I wouldn't want it. Neither would you.'

I said, 'She saw her own end. She knew I'd fail her.'

'What do you mean?'

'I promised to bring her back. To Britannia. I couldn't do it.'

She shook her head. 'Will you be cross with me?'

'Why?'

She said, 'You did bring her back. In your heart. Now you can set her free.'

I said, 'This is a strange conversation.'

She leaned back. She was watching me steadily. She said, 'Have some more to drink.'

'Domina, it's very late. I ought to leave.'

'Late, early, pooh,' she said. 'I only wake up at night. Have some more wine. To please me.'

I knew I should call for my horse, ride back. There was work to be done, too much work; the morning would come too soon.

'Yes,' I said, 'thank you. I'll have some more wine.'

The liquid made a little tinkling sound. The bowl had been emptied, refilled, emptied and filled again. She seemed unaware. She said, 'Are you still angry?'

'No.'

She laughed. 'Did you know you had a nickname?'

'What is it?'

She said, 'They call you the man who never smiles. Is it true?'

I said, 'Maybe I don't often find much to smile at.'

She sighed. She said, 'Sometimes you remind me of my father. It's silly.' She paused. 'When I was little I was terrified of him. He had awful rages. You looked so black the other night. As if you could have killed me. When I said about the mines.'

I said, 'I don't want to talk about that now.'

'No,' she said. 'I promised, didn't I? No politics.'

'That wasn't politics.'

There was a silence. I broke it.

'What was your father like?'

She said curtly, 'Like a barbarian.'

'Of course.'

She said, 'Now you're upset again. Please don't be. I'm a queer old thing. Prickly. You're prickly too.'

I said, 'I'm not upset. What was your home like?'

Her eyes were cloudy. 'There was a tower,' she said. 'On an island, by the sea. And a stockade where we drove the animals. And beyond, the sea again. Blueness and blueness, for ever.'

'And beyond the sea?'

'Nobody knows,' she said. 'Nobody can go beyond the sea.'

'Except the Boat.'

'It was my Boat,' she said. 'I saw it too. White, like a swan.'

I said, 'You believe in the Land. Very much.'

'Yes,' she said. 'Don't you?'

There was a night bird, with a bubbling cry. It sounded clearly in the quiet. I said, 'I must go.'

'No,' she said. 'Not yet. Please.' She sat up. 'Will you do something for me?'

'What?'

'Come and meet the Nymph.'

'Where is she?'

'At the temple. It's full moon. She always comes at full moon. She's sad because nobody worships her any more. But it's her temple, it was built for her. She'll never leave.'

I said, 'Won't the servants think it odd?'

'There aren't any servants. Not in this wing. Except my own. And I always go there, anyway.'

The wine had made me sleepy. I made two attempts to rise. She gripped my arm and laughed. The touch felt strange.

The moon was high, riding a serene sky. The villa buildings seemed to loom from a greenish-silver haze. It was like a dream, or walking under the sea.

She moved ahead of me. Her sandals lisped against stone. Beyond the bath-houses a shallow flight of steps led upward. The shafts of the temple gleamed like tall bones. The shadows they cast were velvet-black.

There was a little pool, surrounded by a low stone coping. She sat dabbling her hand. She seemed to watch me, but the moonlight made her blind. The bird called again, close overhead; the night breeze moved in the spinney, whispering the branches together. This, I thought, would make a fine tale for Valerius. I set my mouth.

She said, 'Can you see her?'

'No.'

She laughed. 'You should have brought your wine.'

My head was spinning enough. I moved away a little and turned. I said, 'I see her clearly now. She's very lovely.'

She shook her head. She said, 'She couldn't come tonight. There's only me. I'm not lovely.'

I said, 'She wouldn't come.'

'Why not?'

'She was offended. She knows her water runs to the latrines.'

She frowned. 'Sergius, why didn't you ever marry?'

'I never had the opportunity.'

'I'm sure you had loads of opportunity. Didn't you ever find a woman you could love?'

I said, 'They didn't love me.'

'That can't be true.'

'I didn't have enough to offer.'

She said, 'I'd say you have a lot to offer. I don't think they knew what they were missing.'

I was silent.

She said, 'Why are you so bitter?'

'I'm not bitter. Just a realist.'

'You're bitter. You'll fall in love one day.'

'No.'

'You'll find a woman who wants you. She'll make it happen.'

'I don't think so.'

'I know so.'

I said, 'There's always a latrine.' I rose, unsteadily. 'I'm sorry, I think I'm a little drunk.'

'I don't care. So am I. Let's fetch some more wine.'

'I don't want any more.'

'I do. I want to drink and drink. All night.'

I didn't answer.

She said, 'Please sit down.'

I sat.

'What are you thinking about?'

'Nothing. The night.'

'What about the night?'

'Not wanting it to end.'

She said quietly, 'Why don't you want it to end?'

I thought before I answered. I said 'Because nothing like it will come again.'

And that, I thought, is the result of too much booze. I hadn't been serious about making a fool of myself twice. Now I was glad Valerius hadn't come.

She said, 'I don't want it to end either. But you don't believe that, do you? You're a realist.' She waited, but I didn't comment. She said, 'Where do you live? In Corinium?'

'Yes.'

'What's it like?'

'A scruffy little room. Bedroom attached.'

She drew her legs beneath her, clasped her hands round her knees. 'Why do you think no woman will love you?'

'It's what I believe.'

'You don't love them.'

'Perhaps.'

'How did you hurt your hand? Was that to do with a woman?'

'No.'

She said, 'If the Nymph came, she'd make you love her.'
'I don't think so.'
'You wouldn't have any choice.'
'I would.'
'Be careful,' she said. 'You're mocking the Gods.'
I felt an odd, hollow anger. I said, 'Then let them be mocked. . . .'
'No,' she said. 'Sergius, no . . .'
'Very well. If it displeases you.'
She licked her mouth. 'If I loved you, would it teach you differently?'
'No.'
'Why not?'
'Because you don't,' I said. 'And can't.'
She rose and walked towards me. Her face looked frozen, like stone. She sat, put her arms on my shoulders and drew me towards her. She opened her mouth, pressed my lips to hers and pushed as deep and as far as her tongue would go. Then she took my hand. 'Touch me,' she said. 'Please touch me.'
I'd like to say the earth reeled. It didn't. Nothing much happened at all. I was too surprised.

We lay in darkness. At first, shock had kept me limp. Now she moved against me, and my body rose. 'Feel me,' she said. 'Feel what you've done.' I kissed her lips; and there was a second kiss, deeper and sweet. I had wondered, absurdly, if there might be pain. There was none. Just the smooth warmth. Afterwards I lay and laughed, feeling the sweat between us and the sliding of her breasts. I said, 'I was a virgin.'
She said, 'You're not a virgin now.'
'Crearwy . . .'
'Hush,' she said. 'Rest. Sleep now.'

When I woke there was grey daylight in the room. A solitary bird was piping in the woods. At first I couldn't remember where I was. Then I reached across the couch. It was empty.
There was a scraping at the door. I sat up alarmed. It was the girl Pelgea. She set down a pitcher of water and a bowl. I smiled at her uncertainly. She slipped back through the door and was gone.
Oddly, I didn't want to wash. It seemed I was laving some

part of her presence from me. When I'd finished I found my tunic and dressed. I stepped outside. The sun was just up. The lawns were grey with dew; across them lay the long, angled shadows of the buildings.

She was waiting for me. The triclinium looked strange in the early light. There was bread and fruit, milk, well-watered wine. A bowl of honey. I ate awkwardly, in silence. Then I looked up. I said, 'Crearwy . . .'

She smiled. She said, 'Was it nice?' Her voice sounded small.

I said, 'I didn't want to wash. It was strange.'

She said, 'I know.' She came to me. We kissed. This time there was warmth. She pulled away. She said, 'Oh, no. No, no . . .'

'Crearwy . . .'

'Yes?'

I said, 'The Nymph came. After all.'

She said, 'I didn't know it was going to happen. You made it happen.'

'I?'

'You shouldn't look at a woman like that.'

'Like what?'

'That first night. It was terrible. It went right through me.'

'What?'

She said, 'Didn't you know? It was your eyes. They get bluer.'

'So do yours.'

She broke a piece of bread, smeared it thickly with honey. She said, 'I'm sure they don't.'

'Have some bread with your honey.'

She started to laugh. Her mouth was full and she coughed. She said, 'It's ridiculous.'

'What is?'

'Us. Sitting like this. Talking. No, not ridiculous. Strange. Nice.'

I said, 'I shall call you the Honey Princess. It's your hair colour.'

'Mmm. That's nice. Sergius . . .'

'Yes?'

'Are you angry with me?'

'Should I be?'

She said, 'I thought you'd be . . . disgusted. You wouldn't want to see me again.'

'I'm not disgusted.'

I pushed my plate away. She said, 'What's wrong?'

'Nothing. I don't want you in trouble.'

'I shan't be in trouble. I shall be all right.'

'How do you know?'

'I'm always all right.'

I said, 'I must go now.'

'Not yet. In a little while.'

'I must.'

'Your horse isn't saddled,' she said. 'Walk up to the temple with me.'

The pool was eight-sided and shallow. Its water was crystal clear. She leaned over it. She said softly, 'Thank you, nymph.'

There was something scratched roughly on the sill. I hadn't noticed it the night before. The Greek letters Chi and Rho, for the Christos. I touched the mark vaguely with my finger. I said, 'Are you a Christian?'

'I believe it's all the same God. I don't know what I am.'

'But you believe in him.'

'Yes,' she said. 'Yes.'

'When shall I see you again?'

She smiled, watching back at me; and I saw something I was to notice many times, how her eyes were never quite still. They moved constantly, in little shifts and changes of direction; yet all the time stayed fixed on my face. Many lovely women seem to have that gift.

She said, 'Will you come soon?'

'Yes.'

'How soon?'

'When I've made a bow.'

She walked with me to the gate. The villa was awake now and bustling. Pails clanked and rattled; there was a noise of hammering, somewhere a man sang lustily. The place sounded like what it was; a great farm. I turned back. The lines of windows seemed to be staring. I said, 'If you write, use better messengers.'

'Why?'

'Felix reads your letters.'

She coloured instantly. 'How do you know?'

'I know.'

She bit her lip. 'I'll be careful. But I can't write much.'

'I know.'

My horse was waiting, a slave holding its head. I mounted. I said, 'Thank you for a pleasant stay, Domina. My regards to your husband, if he's back before I come again.'

She stared up at me. She said, 'Have a safe journey, Praefect.'

'Thank you. Goodbye.'

'Goodbye.'

The hooves echoed under the arch. Then the place was behind me. I turned once. She was still standing by the gateway. We didn't wave.

I let the horse walk, feeling the sun warm on my neck. I had entered a new state, a state I as yet barely understood. Yesterday the future had stretched as barren as the past. Now it was confused and rich. Yesterday my life had been my own. Now it was divided. It came to me, not for the first time, that a man is the sum of what he knows, that as the total of experience inexorably mounts so the breathing creature alters, is reborn. Yesterday I had been a virgin. Now I was not. I searched my awareness for the sense of sin. I found nothing. Our coming together had been a Fact; as the green grass round me was a Fact, and the bushes and waving trees. Parts, all, of the great Fact that is It, the unknowable, Oneness. Lost in the welter of Being, right and wrong lose coherence. I can neither excuse nor justify in the conventional terms of faith. Let the pure among you, blind things acting blindly, cast your stones.

I left the track where a wood swept down to meet it. It was cool beneath the trees. There was a streamlet, chuckling among boulders. In one place it had swelled to form a little pool. The water was clear and cold, like the pool in the temple. I sat and watched the weed move slowly, like graceful green hair. A close sound startled me. I turned. The horse was drinking noisily.

I rode on. Eventually I came to a clearing. There was a stockade, with a round thatched hut. Children played in the doorway. I called, and a woman answered. I knew a few words of the local tongue. I asked her for something to drink. She brought a bowl of milk. It was all they had.

I sat a while in the shade. I had lost my way. When I rose I said, 'Corinium?'

She pointed. Corinium, yes. There. Corinium. There . . . I gave her a piece of gold. She stood staring at it, biting it wonderingly.

The walls of the town were bright in sunlight, striped with their long bonding courses of red brick. The sentries at the gate saluted me. I rode through. It was market day, the streets were crowded. I stabled the horse, walked to my office by the basilica. Valerius met me. He stared curiously. He said, 'Did it go off all right, sir?'

'Thank you,' I said, 'yes.'

'We were expecting you last night.'

'It got late.'

I slung my sword and baldric on the bed. My desk looked cluttered. I sat down. He tapped the door behind me. Little noises still tended, obscurely, to make me jump.

'Come.'

His hands were full of papers. He said, 'Despatch from Com. Lit. Sax., sir.'

'Who?'

'Count Hnaufridus.'

I said vaguely, 'Must you use that horrible slang? What's the other?'

'From Camulodunum.'

'Thanks. I'll look at them later. Right now I want to bathe.'

'Yes, sir. By the way . . .'

'Hmm?'

'Any idea when we're leaving for Isca?'

'Not yet a while. I may leave it over. Valerius . . .'

'Sir?'

'Can you lay hands on a good bowyer?'

If he was surprised he didn't show it. He said, 'I expect so, sir.'

'Good. Get him to come along this afternoon, if you can.'

'Very good sir. It shall be done.'

Insensibly, in my memory, the season alters. The leaves deepen from their first translucent green. Oaks hang out their massive clouds of foliage. The barley ripens in the little square fields. The thin spears of spring are gone; now the grain stands heavy and tall. Soon it will turn golden.

Nessa rides to meet me, across the field. She's whooping,

holding up a freshly killed rabbit. Her arrow still transfixes it. Blood from the trophy has run back down her arm. In her triumph she is unaware of it.

Crearwy said, 'You love them, don't you?'
'Who?'
'The children.'
'No.'
'I think you do.'
'I don't.'
'Why not?'
'They're not mine to love. I can't afford it.'

I was lying, of course, to suit my own queer humour. Those long-haired rats, shinning up apple trees, showing their funny little drawers, delighted me. But love is a two-sided coin. Sorrow must balance joy.

We were sitting in the little temple. The warmth had made me drowsy. There was a scent of thyme and lavender and the midsummer hum was loud.

She said, 'I love this place.'
I didn't answer.
She said, 'There's peace here. This is where the bees come to die.'
I said, 'Why did you marry him?'
She had been plaiting stems of lavender. She put them down. She said, 'I don't know.'
'Did you love him?'
'He was good to me.'
'Isn't he now?'
'Yes,' she said. 'In his way. Sometimes. I don't know. He's kind.' She looked up. 'There was another child,' she said. 'The first. It died.'
I waited.
'He wasn't here,' she said. 'He went to Gaul.'
'Till the fuss blew over?'
'Something like that.'
'That sounds a kindly act.'
She said, 'He gave me a roof.'
I turned to look across the green expanse of lawns. I said, 'Such a roof.'
She said, 'He's coming home tomorrow, by the way. He's in Britannia again. I had a letter.'

I got up. I said, 'Thanks for letting me know.'

'Don't be angry. I couldn't help him coming back.'

'You could help being so bloody casual. Couldn't you have told me?'

'I did tell you.'

'Yes. When you had to.'

She bit her lip. She said, 'I didn't want to spoil what we had.'

'Meaning we'll have no more.'

'I couldn't help it. Please don't go.'

'I'd better. He might be back sooner than you think.'

'He won't be. He always stays a few nights in Dubris.'

'Excellent. I'm glad you all work to a system.'

'You are in a bad temper,' she said. 'It wouldn't matter, anyway, if you were here.'

'Maybe not to you.'

She flared at me. 'You're not the only one with pride!'

'From where I'm standing,' I said, 'it looks as if I might be.'

Her face was like an animal in pain. When she was hurt she always looked like that. I wanted to take her in my arms; but also, obscurely, I wanted to wound her again. Her offhandedness infuriated me. If that was all our relationship meant it was better ended quickly. I turned to walk away.

She said, 'Where are you going?'

'Does it matter?'

'It matters to me. You know it does.'

'It sounds like it.'

She put her face in her hands. She said, 'Anyway, it wasn't your business.'

'What wasn't my business?'

'When he came back. I can be nasty too.'

I said, 'You certainly can. Thanks for a pleasant summer.'

'There's no need to insult me!'

'I didn't insult you.'

'You damned well did!'

'Well, it makes a change,' I said. 'Usually it comes my way.'

She stared at me, tear-stained and angry. 'Sergius,' she said, 'please don't let's quarrel. It's been so lovely. . . . Oh God,' she said, 'what's going to happen to us?'

'What should happen to us?'

'Do you expect me to run away with you or something?'

'Any other man might. Not me.'

'I couldn't,' she said. 'I couldn't. If you had children, you'd know.'

'That,' I said, 'is the hammer I was waiting for you to use.'

'What hammer! I don't understand!'

'You wouldn't.'

'The children come first. They always come first. I always put them first.'

'Except once.'

'When was that?'

'The night you seduced me.'

'*I didn't seduce you!*'

'I can't think of a better word for it right now.'

She said, 'You'd better go. No, please don't.' The tears were splashing again, unchecked. 'I don't know what I want,' she said. 'Do you think I ever pleaded with anybody before?'

'I don't know what you did before.'

She scrubbed angrily at her face. She said, 'It's only the children.'

'And the rest.'

'What rest?'

'A cosy little villa.'

She said furiously, 'Do you think that matters?'

'It mattered when you married him for it.'

'I didn't marry him for it. If you want to know, he bought me.'

'There's two sides to that,' I said. 'He bought. You sold.'

'You never loved me. Ever.'

'I always loved you. I still do. But I feel out of place.'

'You're not out of place! Please stay. I don't want you to go.'

'I don't particularly want to meet your husband.'

'Why? Is your conscience troubling you?'

'My conscience is clear. I'm just tired of his manners.'

'There's nothing wrong with his manners!'

'There was plenty wrong the first night I came here.'

'Oh, that,' she said. 'Why do you keep going on about that? It was only a joke. Oh, believe what you like. I don't care. You won't believe the truth.'

'I saw the truth. You rather wanted to show off the British standard of living. I admit it's higher than mine.'

She muttered, 'It needn't have been.'

'What was that?'

279

'Nothing. Forget it. Just go.'
'I asked you what you meant.'

She rose slowly to face me. Her voice was low and clear. 'The first night you came here,' she said, 'you were offered the Purple. *And you turned it down.*'

Chapter seventeen

I rode to Isca. I spent a month in the town. My days were tiring, but I found I couldn't sleep. I took to drinking more wine than was good for me. In time I sickened of it. I still slept badly.

I'd expected the pain to ease by degrees. It didn't. I lived a second shadowy life. I saw her and heard her, in the quiet watches; her laughter, the turn of her head and ankle, the flash of her eyes. If she'd laid some Island curse on me it couldn't have been worse. I saw the villa and the temple, the white gleam of its columns. Later I saw her body. She put the fillet from her hair, slipped the robe down low round her hips. Her waist was slender as a statue, her belly full and not like a girl's. Her nipples jutted boldly, pointed and pink. She shaved her armpits but not her body; her loin hair was harsh and dark and curly. Sometimes after we had made love she used to sing; a little nonsense song, in a lisping voice. She would never tell me where she learned it, or what it meant.

I talked to her. Sometimes angrily, sometimes with affection. She answered me with words that she had used.

'Sergius, why do you pull things down on your head? Why can't you just be quiet, and enjoy? You make bad things happen to you, if you expect them all the time. . . .'

I said, 'I didn't make this one happen, Domina. . . .'

Other times the sweat stood on my body. I needed to slide into her, unroll like a glove.

I wrote her letters. I always tore them up.

I'd been offered accommodation at the house of a curiale. I refused it. Valerius had brought a tent. I had it pitched in an open space under the town wall. We were building catapults. I slept on the job.

The days were filled with noise. In the mornings I'd oversee the training of the new militia. They slashed and banged at the practice stakes, charged up and down their makeshift parade ground uttering fearsome cries.

'No,' I'd say. 'Like this. No, this way. Keep your shield-rim up, damn and blast you. Cover yourself, man, cover and thrust. *Thrust*, don't dangle it like a bunch of wet lettuce. *Thrust*, damn you, *thrust*. . . .'

In the end I fought a practice bout with Valerius. He came at me with unexpected vigour, forced me to give ground. He knew some unpleasant moves. He offered a hamstring cut. If he'd followed through he'd have reached me. I ducked my shoulder and barged him, slamming with my shield. He measured his length and I stood over him. I swung my arm and let it fall.

He was watching up at me with an odd expression. He said, 'Permission to get up, sir?' His voice sounded strained.

I grunted and stepped aside. He rose gingerly, fingering his temple. There was a red weal where the shield-rim had caught him. He rubbed it. I said, 'Sorry about that, Tribune, but you asked for it. Next time be more careful. Fancy strokes are all very well, but be ready for the counter. Otherwise you'll find yourself suddenly dead.' I wiped my face. 'I'll buy you a drink,' I said. 'You've earned it.'

We sat in a tavern. I found my hands were shaking. I'd strained my finger, too, in the fight. It was starting to throb. I must be getting old.

He was still watching me. He said, 'I thought I'd bought it for a minute then.'

'If I'd been Scotic, you would have done,' I said. 'It's not a game.'

He said, 'You're a bloody good swordsman, sir, if you don't mind my saying so. Where were you trained?'

'In Hispania,' I said. 'When I was a lot younger than you. At least it's one thing I can do.' I glanced across at him. 'Where do you come from, Valerius? You've never mentioned your people.'

'Oh,' he said, 'they were nothing much. They died when I was young. I lived with my aunt. I was glad to get away.'

'Where was that?'

'Brigantia.'

'What?' I said, half to myself. 'You too?'

'Sir . . .'

'Yes?'

He looked embarrassed. 'I'm . . . can I make a comment?'

I said sourly, 'Be my guest.'

He said, 'You're not . . . well, you don't seem your usual self. Is there anything I can do?'

I stared at him. 'I don't think so, Tribune,' I said. 'Thanks anyway.' I drained my cup. I said, 'I think we'll go back.'

'Back?'

'To Corinium. I'm sick of this collection of amateur gladiators.'

'Do you think they're good enough to leave?'

'No,' I said, 'I don't. One look at an enemy and they'll probably all run like Hell. But no amount of prodding's going to alter that. And to be quite frank, I don't give a damn.'

We rode through Lindinis and Aquae Sulis. I'd felt vaguely curious to see the latter town. The baths are the thing, of course. They say the healing properties of the waters are remarkable. I was impressed despite my mood. Rain was falling, dimpling the great dull-green pools. Steam blew steadily across the surface of the water. The spring that feeds the place comes roaring up from a grotto, piping hot. I listened to the bubbling and splashing and thought, She'd like it here. If ever a cave was haunted by a Nymph, it was that one. I was still thinking of life in terms of Crearwy. Once she had been Coventina. Now she was Minerva-Sul.

Corinium was much as I had left it: noisy, and a little malodorous. I rode in grimly through the gates, Valerius at my heels. I'd achieved nothing by coming back. I was merely nearer the source of pain.

There were letters waiting for me. Tammonius had written, irately, from the north. What the Devil was going on in my sector, where were my reports? He told me the Picti had been active again along the entire Wall. He had had to reduce the flank garrison at Deva. It all seemed very far away.

There was a wooden tablet. I left it till last. I didn't want to open it. Eventually I cracked the seal.

You anger me almost as much as I love you, she had written. *When will you come to me again? Forgive me, Sergius, for what I said. I want you so badly. I love you, I love you. My soul cries out for you. Does that sound stupid? I'm sorry, it's what I feel. I haven't your use of words. See, you've made me speak. Women aren't supposed to say these things. Haven't you heard me calling? I was a bird, that flew over your roof. I was a cat, that watched you through the window. I was a fox, I saw you ride along the road. Nothing matters except that you come.*

Please trust me. Can't you see how much you mean to me?

There was a strong-box in the room. I locked the letter carefully away. Before I slept I spent an hour trying to resurrect the remains of my pride. In the morning, at last, there was peace.

'See,' I said to Melinda. 'When you pull the string, the monkey climbs the stick.'

She stared, intrigued. 'What has Nessa got?'

'A wooden lion. He can wag his tail.'

'Monkey's better,' she said. 'He climbs a stick.'

When they had gone I said, 'Parcels for you, ma'am.'

She said, 'Oh, no . . .'

I'd wrapped them as Calgaca used to, with sprays of green leaves on top. She stood them on the table and walked round them. She said, 'I don't want to open them. They look too nice. I'm being a cat. Sniffing them all over first.'

'Are you a cat?'

She mewed, and pounced.

She slipped the bracelet on her arm. She said, 'It's Scotic.'

'I know.'

'What's in the other?'

'Walk round it some more.'

'I can't tell,' she said. 'It's heavy. It isn't wine.'

'No. Not wine.'

She undid the wrapping, gingerly. Lifted the crock, and sniffed. She said, 'It's honey. It can't be heather honey.'

'From Dalriada. A piece of your sky.'

She wiped her face. She said, 'I'm such a fool.'

'Here!'

'I cry when I'm happy,' she said. 'I cry when I'm sad. Sergius, you are good to me.'

I hadn't expected her to be moved. But she was homesick; and bees, as is well known, gather an essence from the ether through which they fly.

Later she said, 'Sergius, I love you in so many ways. Did you want me?'
'Very much.'
'Do you want me now?'
'I don't know.'
'Shall we go to bed?'
'No.'
'What's wrong?'
'Nothing. I just don't want to.'
She said, 'It's the children, isn't it?'
'Yes.'
'They were here last time.'
'I know.'
She said, 'You're right, of course. Do you want to go?'
'No.'
'What shall we do?'
'Nothing. You can invite me to dinner. Pelgea can sing.'
'Can we talk? I just want to be with you. All the time.'
'Yes,' I said. 'We can talk.'
'That will be nice. . . .'

She said, 'You've made a hit with Melinda at least.'
'Hmm?'
'Melinda. She's mad about you.'
'I can't see why.'
She was reclining, as she usually did when we ate alone. She pushed her plate away and smiled. 'You can never see. You always underrate yourself.'
'What does she say?'
'Asks when Sergius is coming back.' She paused. 'She knows I love you.'
'She's too young.'
'She's not. I heard her telling Nessa.'
'Good God! What did she say?'
'Nothing. She's very reserved.'
'She takes after her father.'
'Perhaps.'
A silence; then she sighed. 'I went to see them last night,' she said. 'After they were asleep. I think the lamp disturbed them. Melinda said, "Mummy, I do love you." She didn't wake.' She smiled. 'What can you do?'

'Nothing. It's our tragedy.'

She reached to touch my wrist. She said, 'It was such a long time.'

'I know.'

'Were you unhappy?'

'I nearly killed Valerius.'

'Who?'

'My Tribune. Don't you remember him?'

'Of course. Now you say. He's nice.'

'He's a good man.'

'I think he loves you too. In his way.'

I was silent. She said, 'What else did you do?'

'Made some catapults. Not very interesting.'

She said indignantly, 'I think they're terribly interesting. I always wanted to fire one. Wham great rocks about.'

'That's strange. For a girl.'

'I didn't want to be a girl. I wanted to be a boy. All my father's other children were sons. They used to take me fishing. I wanted to go with the war boats. They would never let me.'

'I'm glad.'

'Why?'

'You might have been killed.'

She said, 'Nobody ever cared about me before. Whether I was killed or not.'

'Didn't your father?'

'He sold me. Girls aren't any use. Censorinus had me educated. I was ten.'

I sat up. I said, 'That's the most horrible damned thing I ever heard.'

'It didn't matter.'

'Crearwy . . .'

'Yes?'

'What did you mean? About the Purple?'

'Nothing. I told you. I was being silly.'

'I want to know.'

She frowned, staring down at the table, tracing a pattern aimlessly with one finger. She said, 'It's . . . I don't know. Really I don't. He doesn't say much to me.'

I waited.

'He feels . . . very strongly,' she said. 'About Rome. And Britannia.'

'Who was the other man? Gratianus?'

'I don't know.'

I said, 'You wouldn't really want to see me beheaded, would you?'

She looked at me sombrely. She said, 'You'll make me miserable.'

'I feel strongly too.'

'Yes,' she said wryly. 'I remember.' She frowned. 'Sergius...'

'Yes?'

'Why were you angry? About the mines?'

I said shortly, 'I ran one.'

She said, 'Was it very bad?'

I rubbed my face. 'How old's Nessa?'

'Ten. Nearly eleven.'

'Do you know what she'd be doing? If she was a mine slave?'

'No.'

'Working at the face. Or on the wheels.'

'What are the wheels?'

'They lift the water. Like a treadmill. She'd be naked. If she wanted to relieve herself she'd have to do it. The wheels don't stop.'

'Sergius...'

I sat up, and poured myself some wine. 'She'd have sores,' I said. 'On her feet and shoulders. Probably inside her mouth. She wouldn't be a virgin, of course. Not at eleven.'

She said, 'You were badly hurt.'

'I was the overseer. It was the slaves who were hurt.'

She stared at me. She said, 'I didn't know.'

'You weren't expected to.'

She said, 'That's why you were angry. Because of how we live.'

I didn't answer.

'They were left to him,' she said. 'The mine shares. He doesn't make much profit on them.'

'That's bad luck.'

'You think we're hypocrites.'

'I think you're fortunate.'

'Why?'

'Your children see the sun.'

She said bitterly, 'Would you prefer to see them in the mines?'

'No,' I said. 'I wouldn't want them to be in the mines.'
A pause. She said, 'Have something else to eat.'
'Not for the moment, thank you.'
'Sergius . . .'
'Yes?'
'I'm sorry.'
I shrugged. 'Where's your husband?'
She looked at me warily. 'He went back to Gaul. I don't know why. He was only here two days. Sergius, promise me something.'
'What?'
'Don't get us into trouble.'
'How could I?'
She said, 'You're an Imperial Praefect.'
'Not a real one.'
'Why not?'
'Real ones have Legions.'
She gestured, silently, to Pelgea. She came forward, with the harp.

I could have stayed overnight. I refused. It was dangerous for her. She came to the gate with me. There was a groom with a torch. I walked my horse out of earshot. She said quietly, 'When shall I see you?'
'I don't know.'
'Will you come here?'
I shook my head. 'It feels wrong.'
'Would it feel righter anywhere else?'
'It might.'
She said. 'I'll think of something.'
'What?'
'Just something. I'll write to you.'
'Be careful.'
'I will. Sergius . . .'
'Yes?'
'Nothing. I'm being silly. Ride now. You'll get cold.'
I said, 'Good night, Crearwy.'
'Good night. God bless.'
I was cold before I reached Corinium; cold in my bones. The journey back seemed endless. As I neared the town I remembered I hadn't told her about Aquae Sulis.

The children are away with friends, she wrote. *They usually go about this time of year. C. has written from Gallia. He'll be back in a month's time.*

I'm afraid I feel a sickness coming on. I shall stay in my room. Pelgea will look after me. But I really think sea air would do me good. Do you know anywhere nice? Think of it, Sergius; us, on our own!

I cried when you had gone. I wanted to run after you. Please write to me. I'm so lonely without you.

I feel as if I've known you years. As if I've always known you. . . .

I sent for Valerius. When he came in I mixed him a glass of wine. He sat sipping solemnly, watching my face.

'Tribune,' I said abruptly, 'when we were in Isca you asked if you could help me. Does the offer still stand?'

'Yes, sir,' he said. 'It does.'

'Don't misunderstand me. This a matter between friends.'

'That's all right sir,' he said. 'You've been good to me since you came to the Province. You can count on me.'

'I shall have to,' I said. 'By the Gods, I certainly shall.' I outlined my requirement; a quiet place, preferably near the sea, to which I could retreat for a few days at a time. When I'd finished he said briskly, 'I think that can be arranged, sir. When were you thinking of leaving?'

I stared at him. 'Valerius,' I said, 'does nothing ever surprise you?'

'Frequently, sir,' he said. 'But not this.'

'Why not this?'

He hesitated, and took the plunge.

'She's a very lovely lady, sir,' he said. 'I think you're most fortunate.'

'Fortunate? That's a strange way to look at it.'

He said, 'Maybe so.'

I shrugged. 'It seems we understand each other better than I realised.'

'Yes, sir.'

'Had you anything specific in mind?'

'As a matter of fact, I have,' he said. 'I shall need a couple of days' leave though. I shall have to cross the Sabrina.'

'You come and go as you please. You know that already. Where is this place, in Siluria?'

'No,' he said, 'it's not. It's down south. Bit of a ride, I'm afraid. But as it happens, it's slap on the coast. I think you'll find it quiet.'

'What is it, a farmhouse?'

'No,' he said. 'A temple.'

'A what?'

He laughed. 'A temple. To my God. Nuada the Hunter. They built it a few years ago. It's very small. There's a priest's house but no priest. He hasn't much of a following in those parts. They'll be glad of somebody to look after it for a bit.'

I said, 'I think I've heard everything now. What will my duties be?'

'I shouldn't think there'd be any. If anybody came it would only be to make a sacrifice. You'd have to receive them; but they wouldn't stay. There's no accommodation for pilgrims.'

'Would not the God object?'

'No, sir,' he said firmly. 'Not if you're sincere. He's a Hunter; but he's also a Physician. He sees into people's hearts.'

I said gravely, 'I'll make him an offering.'

'Yes, sir,' he said. 'I'd appreciate it if you would. Flesh and fruit are acceptable to him, or any produce of the fields.'

'I'll see it's done. Valerius . . .'

'Sir?'

'Was it so very plain?'

He said, 'If you looked.'

'And you don't condemn me?'

'It's against our faith,' he said. 'The Hunter will do the judging, in his good time. It would be premature for us to start now. That's what we're taught, who follow him.'

So it came about, a few days later, that a young Celtic prince left Censorina for the south. He travelled, dramatically enough, at dead of night. He wore a heavy hooded cloak; his hair, escaping from the cowl, flapped long and pale in the moonlight. His saddle trappings were rich; at his side hung a dagger and a short sword. The sheaths clinked musically as he rode.

I hadn't imagined her in tunic and trews. At first I couldn't take my eyes off her. She said, 'Don't look like that or I shall have to stop right now. We've got an awful way to go.'

It was late the following day before we reached our destination. I enquired for the temple at Vindoclavia, turned west along the coast. Folk glanced curiously at the muffled figure beside me but no comment was made. We finally came in sight of the place an hour later. The track we had been following had swung inland; now it angled back southward, towards the sea. The building, white and four-square, crowned a tall headland. Next to it was the priest's cell. I rode towards it, across smoothly sloping grass. Beyond the temple the land broke away in a crumbling sweep to a ragged half-moon of beach. A keen wind blew from the sea, lifting our hair. The place was deserted; there were no sounds except the roll and wash of the waves, the high solitary crying of a bird.

Crearwy said, 'It's beautiful.'

The sleeping quarters were simple and bare. I had brought bedding on a pack animal. I unstrapped the bundles, carried them inside. She said, 'I can't believe it. Are we really here?'

I said, 'We are.' I took her in my arms; a moment later she pushed away. 'See to the animals,' she said, 'or we shall never get done.'

A shallow gully, grass-grown, led down to the beach. I unsaddled the horses and picketed them, walked back to the temple. I stepped inside. I had expected an image of the God. There was nothing; just a little altar, inscribed with his name, and two earthenware bowls. In one of them was some withered fruit. I carried them outside and scoured them with sand. We had brought food with us; the pack still lay on the grass. I took some apples, and a slice of venison. I put the bowls back where I had found them and knelt in front of the altar. Nodens, I thought. If you are truly a Physician, grant me understanding. I opened my eyes. Nothing had changed. The sun slanted through the doorway of the shrine; the sea moved gently on the beach. A bird skimmed past, mewing.

She called to me. 'What are you doing?'

I said, 'Asking the God's blessing.'

'Did he give it?'

She came and knelt beside me. She was quiet for a moment; then she looked up. She said, 'He welcomes us.'

'How can you tell?'

'He's a God of my people.'

That is one of the memories of her I love best. Her hair unbound, her eyes big and dark and happy; the bowls beside her, and the simple altar.

She said, 'It's time to eat.'

The place fulfilled all our needs. A spring ran from the cliffs. I brought back a flask of water. It was clear and sweet. Beside the venison there were partridges and a hare, cheese, wine; and a flagon of what she called metheglin, a drink made from herbs and honey. Later we climbed down to the beach. She wore the tunic in which she had travelled, but her feet and legs were bare. I took her, as she had intended, out among the weed and tangled rocks. I have heard men claim the earth can be made to sway. I found this to be true. Afterwards we watched the sunset turn the sea to gold. We loved again, in the dark. It seemed the sweetness of the honey she had drunk was still inside her. I fought to reach it, with my prick. I could think in such terms now, and glory in her still. The other times had been good, but they were nothing to this. She was inventive, well versed in Roman methods. Twice she cried out, her voice not muted; then she cried again. When we had finished she lay and stroked me. 'I love him,' she said. 'Feel, he's still like a rock. I could eat him.'

If my soul was with the Devil, then the bargain was triple-sealed.

Sleep was like the dropping of a black cloth. I woke sometime before dawn. She was fuzzy and warm, and we joined again. Later I said, 'Come down to the beach.'

'Mmm . . . what for?'

'I want to fish.'

'I want to bathe.'

'You'll frighten the fish.'

'Blow your old fish!'

I had brought a net. I cast it from a jutting spur of rock. It seemed the Hunter had been pleased with our prayers. At the second try I landed a fine silver fish. She squeaked with excitement. I grabbed the creature as it flopped and leaped, stunned it and cast again.

She sat beside me, curled in her cloak. The tide surged and sucked, boiling back from the rocks. 'Look,' I said, 'another. Aren't they beauties?'

She sighed. 'I want to show you Dalriada. Dalriada's beautiful.'

'I'm sorry this place isn't.'

'Oh it is, Sergius, it is! But Dalriada's . . . different.'

'You should never have left.'
'If I hadn't, I wouldn't have met you.'
'You didn't have a choice.'
'I did really.'
'You said your father sold you.'
'In a way. He wouldn't have made me go. Not if I didn't want to. He was too proud.'
'Perhaps he needed the money.'
She lifted her chin. 'He didn't,' she said. 'He was King of all the Isles.'
I drew the net. 'We'll go up now,' I said. 'You'll get cold.'
We cleaned the fish and cooked them over a driftwood fire. Afterwards we walked a little, and talked. In the afternoon we rested and made love; in the evening we talked again. It was the pattern of our days.

When I think back to that time it's the small things I remember most. I see her one morning, sitting dressing her hair. She is still naked from the sea. Sunlight from the doorway lies across her arm, the long curve of her back. The light is trapped, glowingly, by the texture of her skin. She bends her head to the pull of the comb; and the shadow of her breast, with its thrusting nipple, shakes against the stonework of the wall. Other times I hear the seethe of the wind through wiry grass, the endless slow mutter of the waves. I see the bright, surprising rise of a white cloud over the hill. I walk the rocks with their fringes of glistening weed; I smell the smoke of our fire, hear her call to me across the beach. Her voice wakes a momentary echo; the echo itself fades into the shouting of a bird.

Three days we passed, four; then it seemed we both knew, without words, that we had come to the end of our time. Her room at the villa lay empty; there was danger there, a tragedy in the making. That she had dared to come at all appalled me now. In the morning we rode back. Our lives were not our own.

I turned to stare up at the headland as we left. She reined beside me. She said, 'You're thinking it's another Last Time.'
'Yes.'
'It won't be. Sergius, let's go now.'
I had been away from Corinium nearly a week. It seemed impossible that the time could have gone so quickly. I settled back into some sort of routine, not thinking too much. I spent

a few days in Isca of the Silures, listening to routine complaints. Scoti were firmly established at several points on the mainland; nothing short of a full-scale expedition seemed likely to dislodge them. When, I was asked, would Britannia's troops be returned to her? My answer to that was routine as well. I didn't know.

In September we rode to the coast again. A year had gone by, flown; autumn was on the Province once more, that unmistakable tang of burning in the air. I was conscious of a heightened and morbid awareness. The spring behind us seemed far-off as the primal Dawn; the winter to come bulked in my mind like the death of light itself. We stayed two days. By then we were both ready to leave. The little sleeping chamber was chill and dank. Its walls streamed with water; mists hung round the headlands, seeming to magnify the fret and whisper of the waves. It was as if in some way the place had withdrawn itself; it belonged now not to us, but to the sea.

She wrote to me, briefly, in October.

C. is home. I can't say much. I think he's suspicious. He asked a lot of questions about you.

Lately he's been strange. He talks to the children but not to me. If I'd given him a son things might have been different. That was all he wanted from me. . . .

Don't write to me. I'll write again when it's safe. Think of me, Sergius. My thoughts are always with you. . . .

I worried over the thing. I couldn't get it out of my mind. What that Alpine glacier of a man might do I couldn't imagine. In the end, unwillingly, I sent for Valerius again. I'd already written a letter for him to carry. I hedged for a while before I handed it to him. He read swiftly and I watched the frown start to form. Then he set his mouth. He said, 'Is this an order, sir?'

'You know it's not.'

He said tonelessly, 'I'd prefer you to make it official.'

I said, 'I can't do that. Valerius, I'm worried for her. There isn't anybody else.'

He rose, stood staring through the window. He said, 'How long will it be?'

'I don't know. I hope . . . not long.'

He squared his shoulders. He said, 'Very well. I'll report to you before I leave.' The door closed behind him, silently. I sat and looked at it and swore. I'd pressed him too hard, and we both knew it. But I had had no choice. For friendship's sake he had chosen to make my affairs his own; now he was caught, as I was caught, in a web of circumstance.

I turned bitterly to the report I was trying to compile. The price was mounting. She had already cost me my peace of mind and a goodly slab of dignity. Now I had lost Valerius.

The days shortened. November brought freezing fogs, December heavy snow. It fell steadily, swathing the town, choking the roads with drifts. Waggons skidded and jammed in the streets; all traffic ground to a halt. It was the worst winter I had known since coming to the Province. Road after road became impassable, blocked by drifts. The snow still fell. The wind skirled endlessly; and the first wolves came down, howling within earshot of the walls.

Some days I spent in my room. Other times I walked to where the towers and battlements of the town glared against the yellow-grey sky. I would climb the rampart, stand staring into the murk. Somewhere out there lay Censorina, blind as Corinium was blind, and dumb. I found it hard to believe the place even existed. It belonged to summer and the light, things half a lifetime away.

I became ill. At first I thought I'd merely taken a cold. Later my face and neck swelled to absurd proportions. I burned and shivered by turns; when I tried to rise I found I could barely stand. I lay staring at the ceiling, hearing the muffled noises from the street. I had a brazier brought in. The fumes half-choked me, but I needed the warmth. I asked Petronius to try to find a doctor. He came back shaking his head. This was Corinium; there was no doctor to be had.

He had been loaned me by the office of the Praeses. He was a freedman, fat, wheedling and unreliable. I bribed him to send his wife in with some meals. The price was high, the food invariably poor. It didn't matter much. I seldom ate it.

I waited for word from Crearwy. At first, obviously, conditions were too bad; she wouldn't risk a courier on the trip, short though the journey was. Later the weather cleared a little. The snow, still hardpacked, froze over, making movement possible. No letter came.

The fever mounted. As always at such times, I was troubled by waking dreams. It became difficult to trust the evidence of my senses. Once Riconus' face swam over me in the gloom. He told me beacons were alight on the roads to the south; the beacons I myself had ordered to be laid. I gave him the acting rank of Praefect, put the town's defences into his hands. His visit, at least, was real; but at other times I spoke to Valerius and Crearwy, who were certainly not in the room. Later it seemed I was back at the temple on the shore. A white gull hung above it. I floated with the bird, tasting the cold freshness of the wind. Cloud shadows slid by on the grass; below me, sharply detailed, were the green-white draperies of the tide. I experienced again the sense of oneness, the mystic unity of all Being; in my exalted state the rubbing of each grass blade, the movement of each grain of sand, seemed invested with a significance at once cosmic and comprehensible. Later the nature of the visions changed. Once I saw Crearwy by the bank of a stream. She was naked, and stepped into the water. It rose slowly to her belly, very cold and clear. Behind her was the black bole of a tree. I glanced up, and woke with a jolt. Its arms were enormous, filling the entire sky.

The swellings reduced themselves. They were succeeded by an infection of my mouth and throat. The new pain was intense. Eating was out of the question; it was all I could do to sip a little wine. I sat muffled in my chair, sometimes awake, sometimes nodding into a doze. Each day was like the next. No message came; and the first doubts began to assail me, gnawing insidiously. She had found other things to occupy her mind; whatever I had thought we owned was gone.

With the realisation came a new and unpleasant awareness. I saw my involvement now not as the stuff of epics but as a blind farce, a senseless groping for what I would never own. I had maybe thought to emancipate myself, move into an adult, significant world. I had not merely failed; I was lower, in my esteem and the esteem of others, than I had been before.

It seemed my mind was infested by adders or worms. The newness and glamour of sensation faded; I saw, not for the first time, how neatly she had engineered the thing, how with every later move she had turned her own strength to account. In her security, her responsibility, she was unassailable and knew it; I was a safe toy, to be played with or discarded as she chose. I heard her voice again, sharp with anger and pride.

'*If you had children of your own, you'd understand.*'
'*There's nothing wrong with my husband's manners!*'

'Oh, believe what you like. Forget it, I don't care. Just go.'
'My father didn't need the money! He was King of all the Isles!'
Yes, and mine had been an inspector of drains.

I muttered and tossed, hating and needing by turns. I composed the speeches with which I would rid myself of her, finally and completely; honed glittering and unrecorded barbs of rhetoric, knowing all the time I lacked the strength, that this final cage I had built for myself was too strong to break. I couldn't go from her, and couldn't stay. I couldn't live, and lacked the will to die. I saw now why the Gods, if they existed, were sparing with their thunderbolts. There is a greater punishment; and it goes on for ever. I raged against her, wanting her. My thoughts became even more confused. I realised just what manner of thing had destroyed the poet Catullus. The notion brought a fresh welling of contempt; for no great verses would spout from my dissolution. I writhed in self-disgust; yet when I finally slept I perforce must dream it was with her. Later another memory came back to plague me. It was the deer again, that triple-damned creature that had somehow burned itself permanently into my brain. I saw the stone fall on to its skull, the bright, sudden blood; and it was sliding, sliding back to its pit of earth. Its eyes were fixed on my face; but now, of course, they were the eyes of Crearwy.

It was too much. Anything was preferable to that. I had no right over her life, no right to wound her, no right to go or stay. And also—here again, warm and weary, the rushing resurgence—had she not proved her love? What sort of sense would it make, to risk what she had risked for a passing whim? I needed to keep faith, with her if with nobody else; for faith was all I owned. I would never, I realised, know more of her than I knew now; or she of me. This perhaps was an inner meaning of love; why men and women strive so hard to join. Physically, they achieve their aim; but their souls, by the nature of things, remain their own, unknowable. Before the One, endlessly, we are each of us alone. Bedding together doesn't change a thing.

Later this hectic inner activity paled to one comfortless thought. Faithful or faithless, courtesan or saint, what difference did it make? We had both been lunatics together; the course we had embarked on could only end in pain. As for Censorinus, how could he not be aware? What servants existed who didn't prattle or couldn't be induced to speak?

Between Crearwy and his wrath I had interposed one strapping young man with a sword; but he could afford to wait. Right was on his side; he could crush me, as and when he chose, with the movement of a finger.

In March, with the brightening of the days, the wolves came closer. The Scoti followed them; and Riconus moved in strength. It was a well-planned operation. There was a battle, or so I heard. Many barbarians fell. I was unmoved. I felt defeated, drained. I lay in my own private Hell. It was a strictly individual place; the Hell of insignificance, and foolishness, and failure.

Chapter eighteen

A letter came. I broke the seals, without surprise. It was from Valerius.
We've had a quiet winter, he wrote. *Not much to report on at all. I watched things pretty closely at first, as you wished. He's a queer bird all right, but there's one thing I'm certain of: he doesn't mean her harm. It's not his way.*
He left again today, for the north. I couldn't find out where. The Domina has been ill; but it's all a bit queer. She kept to her room for a fortnight; the servants weren't admitted, and Pelgea won't talk. She's about again now, but she looks like a ghost.
You must remember Pelgea. I spent most of the winter learning to speak her language; and that wasn't the only thing I learned, I don't mind admitting! I won't give you a blow by blow account, I don't suppose you'd be interested; but I hope you're not considering recalling me, I'd hate to have my head lopped off for mutiny. I don't want to be cut off in my prime . . .
There was more, in the same cheerful vein. I set the thing aside. Other despatches were arriving now. The Wall had been breached, Magnis and Luguvalium had had to fight for their lives. In the east the British Fleet had established a powerful base; elsewhere the story was wearily the same. Refugees by the score, the hundred; burned villages, burned farms, burned towns. Siluria had been devastated, from Segontium to the Deva gap. Segontium itself had fallen.
I rubbed my face. A brazier glowed in the room; as usual, my eyes were smarting. I turned to the report from Hnaufridus. No great defeats, no real victories; just the slow spread of desolation, senseless as a plague. Nothing from Mediolanum,

nothing from Gaul. I put the letters down. So much for Stilicho, Master of the West.

There was a meeting, in the Curia. I attended it, dragged my shaky limbs on to the dais. The Praeses was there, with a camp-following of clerks; and the Bishop of the town, the quinquennales, duovirs and aediles. An armed guard, the guard I had helped train, surrounded the platform. Corinium, I heard, would carry on the fight; the Dobunni at least were loyal to the Emperor and to Roman ideals.

'Yes,' I said. 'Very good.'

The buds were swelling on the trees. I couldn't think of it as spring. There was another letter, from Censorina. I opened it at once, saw Crearwy's sloping, unfeminine hand. The next day I rode to the villa. I felt like a puppet, acting without volition.

Valerius met me. I wondered, vaguely, how he could have matured so fast. Or maybe the change was in my sight. Crearwy was waiting in the triclinium. The children were with her. Melinda ran to me. She had lost another tooth. She was keeping it as an amulet.

I stared at her when they had gone. She was still pale and her eyes looked shadowed. She broke the silence first. She said, 'Have you been ill? You've lost a lot of weight.'

'A little, yes.' I didn't know what else to say.

'Aren't I to be kissed?'

I rose. She said, 'Don't make it a duty.'

I kissed her. She held me, and she was shaking. I said, 'Have you been well?'

'Quite well. Sergius, what's wrong?'

'Nothing. I did think you might be dead.'

'But I'm not.'

'No,' I said. 'You're not.'

This wasn't the woman I'd lived with at the temple. She was very far away.

'Crearwy . . .'

'Yes?'

I stopped. It was useless. How explain the waiting, and the greyness? The anger, quenched and re-sparked, hopings against hope, self-hatings, bright hot shafts of memory? Useless. Unless she understood. None of it could be told.

'How has Valerius been?'

'The children adore him. Sergius . . .'

'What?'

'I know why you sent him,' she said. 'It was sweet. But there was no need.'

I said, 'You think you know why I sent him.'

These were our voices, certainly, acting out stilted lines. Our voices; but I was as far away as she. I said irritably, 'It's ridiculous.'

'What is?'

'Us. Talking like this. Why didn't you write?'

'I couldn't.'

'What's been wrong with you?'

'Nothing. Who said there had been?'

'Valerius wrote to me.'

'It wasn't important. Forget it.'

'It was important to me.'

'I doubt it.'

'You doubt it! For God's sake!'

'There had to be a row, of course,' she said. 'As soon as you came.'

'There isn't any row. I was worried, that was all. How can I help if you won't tell me what's wrong?'

'You! Help!'

'Yes, me. But I was forgetting. It wasn't my business, was it?'

'If you put it that way, no it wasn't!'

I said, 'Here we go again.'

'You only came here to hurt me. I think you hate me.'

'It seems all we can do is hurt each other.'

'I pity the woman who falls in love with you!'

'I thought you had.'

'Yes. So did I.'

'Meaning you've decided differently.'

'Oh, *bugger you*! *Bugger you!*'

I stopped. 'What did you say?'

She was weeping. I gripped her shoulders. I said, 'This is absurd.'

'I didn't start it!'

'Oh yes you did. I'm not taking the blame for that.'

'You started shouting . . .'

'You started making love.'

'Oh God,' she said. 'Won't you ever let it rest?'

'Why should I?'

She blazed at me. 'If you want to know, I had a miscarriage. Now are you satisfied?'

I sat down. I said, 'What?'

'You heard me.' She was snuffling, choking back tears.

'Was it . . .?'

'Who else?'

'You do have a husband.'

She stamped her foot. 'I stopped *sleeping* with him. Because I wanted you. Oh . . .' She shook her head, like a pony in a cloud of flies. 'I waited and waited,' she said. 'I so much wanted you to come. And now . . . this . . . oh God . . .'

I said dully, 'Were you very ill?'

'I filled the room with blood. I nearly died. It didn't matter. It's all right. Just go.'

I sat where I was. I wasn't really thinking about anything.

'I should have died,' she said. 'Then I'd have been out of the way. I tried to. But why should I? I don't *want* to die. . . .'

I took her arms, made her face me. Her eyes were reddened and salty, not Crearwy's eyes. Her breasts shook; wet tracks were on her cheeks, her hair was lank. She stared at me; then she wept a storm. I sat with her till she had finished.

Later she said, 'We must help each other, you see. There's nobody else.'

I said, 'I tried to give you up. I couldn't.'

She said, 'I tried to reach you. When you were ill. I couldn't. It was funny. Like a wall.'

'It was the same with me.'

'Sergius . . .'

'Yes?'

'I love you. You know that, don't you?'

I said, 'It was the others.'

'Which others?'

'When I was sick. It was like a crowd. All shouting.'

'Who were they?'

'I don't know. Just people. Everybody. All the ones who haven't done what we've done.'

She said, 'They don't matter. We mustn't let them.'

'No. It's our own Hell.'

'Strictly private.'

'Of course.'

She smiled. She said, 'I'm better now. I'm sorry I was so queer. I'm always like this. Either up, or down.'
'Crearwy . . .'
'Yes?'
'I won't make love to you again.'
'*What?*'
'Because of what happened.'
She said, 'I never heard anything so absurd.'
'But . . .'
'But what?'
'What happened!'
'Forget what happened. I'll take precautions.'
'What precautions?'
'That's my affair. Don't worry.'

The new campaigning season started hectically.

Tammonius wrote to tell me the Wall had been consolidated. Hard on the heels of the report news came from the south-west, the great peninsula of the Dumnonii. Beacon fires were alight; a battle fleet, the biggest seen in years, was making its way round the toe of the country, headed for Portus Adurni and the Sea of Vectis. King Niall had made his move at last, striking miles from where I had expected him at the heart of southern Britannia. I alerted Riconus, mobilised the town garrison. Corinium sprang into scurrying activity. I sent messengers north and west, into Siluria, ordering the mobilisation of all available men. I saw to my uniform, weapons and armour, spent three hours drafting a letter to Crearwy. After that there was nothing to do except wait.

That night, and the nights that followed, were the longest of my life. I had reported my readiness to Hnaufridus; soon, inevitably, a summons must come. I reflected, sardonically, that my mental turmoil had been pointless; my problems would shortly be resolved by something as rational and decisive as a sword-cut or an arrow. I had never before feared death in battle, mainly because I had never thought of it. I needed peace now, urgently, when no peace was to be had. There was nothing in sight, for Britannia or the world, but suffering and endless war.

The days dragged out, became a week. Still there was nothing. Then, at last, there was a messenger. He came at wild speed, galloping his horse through the streets. The noise he made alarmed the town as effectively as the irruption of a cohort. Before I had finished reading the despatch the Forum was crammed with an anxious mob, all clamouring to know the worst.

I read the words again. At first I hadn't taken them in. For the first time, it seemed, God had seen fit personally to intervene in Britannia's affairs. The messenger, a Tribune from the Count's headquarters, jerked out a confirmation. There had been a great storm. For miles the southern beaches were strewn with corpses. The battle fleet was scattered; Niall of the Nine Hostages, paramount King of Hivernia, was dead.

I grabbed Petronius. 'You. Quick. Out the back way. Take this to the Praeses.' I scribbled on a pad. 'No, first. Get this man something to drink. Have you see the Praefect Riconus?'

'In the square, I think. There's some men.'

'Get them in here. I want a personal escort.' I wrote again. 'Then take this on to the Curiale. Don't let yourself be seen. Jump to it, man!'

Corinium, and most of the Province, went wild. Bonfires burned on watchtowers and in the streets; a service of thanksgiving was attended by half the town; for nights afterwards nobody thought of sleep. In the middle of the uproar another rider arrived, from Augusta. His appearance went unnoticed; that triumph was my own.

I cracked the great seals on the letter, held it to the light.

My impetuous young friend, he had written. *I hear nothing but good of your administration. Can it be that you have matured at last? If so, then all the Gods be praised. . . .*

Two days ago I had the fortune finally to bring King Radagais to a stand. Twelve thousand of his followers are now enrolled in the armies of Rome. His power is broken, and Radagais himself is dead. The place of the encounter is called in your tongue, Faesulae. . . .

The Magister Militum Alaric and his followers I have ordered to Epirus, where they will wait my pleasure. I cannot as yet release the forces of Britannia. The tide is stemmed, not turned; the final battle has yet to be won. Look to your walls, pray to your Gods; but take heart. . . .

There followed the signature of Stilicho, Magister Militum and Guardian of the Two Empires.

There is in Britannia a recurring strength. Rape her, burn her, fight over her as you will; and in the spring her meadows will blossom, her trees spread their new canopy of green.

Crearwy was planting herbs round the little nymphaeum. She called it her Grey Garden; all the shrubs she set had feathery, silvery leaves. She was putting in a lot of work on the grounds; for weeks I never seemed to see her without a trowel or some other such implement in her hand. She liked nothing better than to slop about barefooted, robe tucked up indecorously round her calves, a battered straw hat on her head. She looked fitter now and brown, the dark marks gone from beneath her eyes. The life of the villa flourished. Nessa and Melinda had taken up hunting with the trident. Sometimes I went with them for a day's sport. They were nimble riders; they managed their shaggy ponies adroitly, jinking with the swerving of the hares. I found I was badly out of practice. In time some of my skill returned, but their bag was invariably better than mine. It was a source of recurring satisfaction to them. Nessa's prowess served to heighten her military ambition. She wanted to learn the use of the sword, but in that alone I demurred.

'Please, Sergius . . .'

'No.'

I had just arrived. I was unbuckling my swordbelt; she caught the scabbard, tried to draw the blade. I eased it from her, gently.

'Why not?'

'Girls don't use swords.'

'I don't care!'

I shook my head. 'It's not for you, little maid.'

Crearwy was watching. She said, 'You feel strongly about that too.'

I hefted the baldric, feeling the weight of the weapon, seeing the worn sheen of the leather. I said, 'I'd like to hang it up for good. Know I was never going to draw it again.'

She said, 'The time will come.'

'Perhaps.'

Riconus was now an officer and a gentleman. I'd never seen fit to revoke his rank. It made no difference anyway; Mediolanum had never confirmed it, and he received no extra pay. None the less it pleased him mightily. He had adopted a thoroughly military swagger, taken to wearing cloaks of a variety of gaudy hues. I took him to Censorina, first charging him, if he valued his new status, to watch his language. The evening was a considerable success. The triclinium was full of talk, the singing, lisping chatter of the West. Valerius translated for me, somewhat smugly. I sat dourly, determined, when the opportunity presented, to make an effort to learn my mother tongue. Afterwards Crearwy said, 'You've been holding out on me.'

'What do you mean?'

'Riconus. He's superb.'

'I'm glad he behaved himself. Once he's had a couple of drinks his vocabulary can get a bit strong.'

'How do you know it didn't?'

'Don't rub it in.'

She was working in the garden. She tipped her hat forward elegantly, sat back and regarded me down her nose. 'As a matter of fact I'm considering falling in love with him. You wouldn't mind sharing, would you?'

'You'd better not.'

'Why? What would you do about it?'

'Demote him for a start.'

'You wouldn't dare!'

'Try me and see!'

Stilicho's victory, and the unexpected peace in which the Province basked, heartened Tammonius into releasing, in all, three hundred men; infantry and light cavalry for the most part, with a smart detachment of German lancers. These last I stationed in Glevum; the rest I disposed as equitably as possible through the towns under my immediate command.

A week later Crearwy visited me in Corinium. Valerius drove her in on a shopping expedition. With them was Pelgea, in a new white linen dress. It set off her dark, vivid good looks. Crearwy wore the formal long-sleeved robe and overcloak of a Roman matron; her hair was piled elaborately, and bound by a fillet. She brought a hamper of game, mostly procured by the children, and a flagon of the honey drink. Petronius, considerably impressed, waddled about stertorously, shuffling the room into tidiness. I smiled at Valerius. 'You're looking horribly domesticated. When can I have my staff officer back?'

He grinned, and took the Pictish girl's hand. 'Pelgea wouldn't like it.'

I raised my eyebrows. 'Are you . . . ?'

'Let's say there's an understanding.'

'Congratulations. Petronius, do you think you could bring us in some wine?' I turned to Crearwy. 'How's your husband?'

'Well,' she said. 'He's coming home next week.'

It was like a sudden douche. I'd almost managed to forget the man's existence. Later the others left; Valerius wanted to show his girl the town. I said, 'Crearwy, what's he doing?'

'Who?'

'Censorinus.'

She frowned. 'I don't know.'

'He's away more than he's home.'

'He always has been.'

'I didn't know he had interests in the north.'

She said non-committally, 'He must have, mustn't he?' She looked round the room. She said, 'So this is where all the work gets done.'

I rode with them to the gates, watched the carriage out of sight. Then I walked back. I made a good supper on what she had brought, but I was unexpectedly gloomy. The place just seemed too quiet. Later I picked up a book she'd sent me weeks before, part of the Meditations of Marcus Aurelius. In all the troubles and alarms of the year I hadn't yet had a chance to open it. I read well into the night.

The days lengthened. The weather was flawless, the mornings fresh and clear, the afternoons hot and still. Often I rode to the villa; for its master, after the briefest of appearances, had once more vanished into the unknown. As far as I was concerned, the longer he stayed away the better. Sometimes I took one of the big travelling carriages the family owned, drove Crearwy and the children out for the day. We visited Aquae Sulis and Glevum, where Nessa was most impressed by the German guards. Melinda had contracted a new ambition. She wanted to ride in a real racing chariot. I said, 'I won't promise, but I'll see what I can do.'

Crearwy smiled. She said, 'You'd spoil them completely if you had your way. Their father never paid that much attention to them.'

'He was always too busy with his mine shares.'

She frowned, and laid her hand on my arm; but said no more.

In deep summer in Britannia the sun never really sets. You can see the glow from it all night long, travelling just under the rim of the world. Sometimes we'd take our supper to the nymphaeum; and Pelgea would bring the harp and sing. They were times invested with a special magic; the whole great building quiet, the night air warm, full of the churring of insects, rich with the scents of meadow grass and flowers. It was a remote world, self-contained and perfect; the rest of Britannia, and her troubles, seemed very far away.

We rode south finally to the temple. Everywhere the land looked fat and prosperous. The harvest was coming in; lines of peasants worked methodically across the fields, swinging their sickles into the grain. The little building when we reached it was unchanged. She gathered flowers and sea-thistles from the cliff, left them for the God. Later she said, 'Sergius, are you happy?'

'Yes.'

We were lying together, in the dark, listening to the sea. She moved against me. She said, 'It's been a wonderful year.'

'I know.'

She punched me. 'Come on, you funny old thing. What's worrying you now? Cheer up!'

'I'm cheerful enough.'

'Then prove it!'

She rose before me in the morning. I heard her calling, and stirred. I stepped outside, yawning. The sun was barely up; sea and sky were both a steely grey. The sea was flat and still, empty to the horizon. She stood below me on the beach. I waved; and she shouted again.

'Come and have a swim!'

'In a minute.'

I walked down the little gully, jumped on to the shingle. She stood, back turned to me, and wriggled out of her tunic. She wore nothing underneath. She said, 'Come on.'

'Not now.'

She turned, and stared. 'What on earth have you got that for?'

'It's all right. You bathe. I'll watch.'

The unease wouldn't let me be. I climbed to a rock, sat with the sword across my knees, watching round at the horizon and the sky. The waves lisped and creamed; nothing else moved.

She swam out a long way. Finally I stood and called. She waved, and turned. It seemed an age before she waded from the sea. Her hair, dark with water, trailed across her shoulders and breasts. Her neck and face were brown, her body white as marble. She stood a moment, watching me; then she stooped, picked up a cloth and began to towel herself dry.

It's a final memory.

The day and its brightness slide imperceptibly into night, night to the darkness of a dream. In the dream, faint at first then closer, the rumble of hooves. The noise jars in the ground.

She was gripping my shoulder. 'Sergius . . . Sergius . . . wake up . . .'

I sat up. Fear was a cold hand round my heart. She thrust a tunic at me. I struggled into it, grabbed for my sword, felt the cold roughness of the hilt.

The rumbling stopped. The place was velvet black; outside I could hear the movement of horses, the creak and jingle of harness. A voice called, close and imperious.

I felt a sort of sick grief. We had missed our chance. We could have slipped away into the gully; now the riders had the place surrounded. The voice came again, harsh with authority; and something, some missile, rapped against the woodwork of the door. I said bitterly, 'Barbarians . . .'

Her hand was on my arm. Her voice sounded shaky with relief. 'No,' she said. 'No, *Scoti* . . .' She called, excitedly.

'Crearwy, don't!'

'It's all right!'

I heard her fumble with the catch. The door swung back. I swore, and ran out behind her.

The moon, high and full, touched the headlands with silver. Beyond was the cold crawl of the sea. Between it and the temple, bunched against the brightness, were the riders. Some carried torches. Their glare competed redly with the moonlight. At sight of me the ring tightened menacingly, and a man rode to us across the grass, sat staring down. He was young, not above eighteen or nineteen years. His mane of hair gleamed darkly red. His chest was bare; on his breast hung a massive golden medallion. His face was hard, high-cheekboned and with long, slitted eyes. He wore trews of patterned plaid; a cloak of the same complex weave hung from his shoulders. He sat a massive, rawboned horse. Its reins and harness glinted, encrusted with more gold. The reins he held loosely, bunched in one sinewy hand. In his other hand was a heavy unsheathed sword.

He stared some time before he spoke. Crearwy answered him, quick and low. His ringing voice interrupted her. She listened; then she turned to me.

'He is Ossa,' she said. 'His father was Niall of Hivernia. He is under a vow. The old King swore to conquer Britannia for his people. Now he is King in his place.'

I took a deep breath.

'Tell the barbarian,' I said, 'this land is already claimed. And has been claimed, for the lives of many men. Its master is Honorius, Emperor of the West; and I am his servant. Tell him to take his warriors quickly and go, before the fate that came to his father falls on him.'

He barked a question. She answered. The syllables stumbled and rushed.

'He says your sorcery is strong. It killed his father and two brothers, also many men. But he knows a stronger magic.'

'There is no magic,' I said patiently, 'but the power of swords. This is our magic, and our strength.' I pointed. 'Tell him if he goes to the beach, and counts the grains of sand, by dawn he won't have numbered the armies of the West. Tell him also we are tired; of destruction, and burning, and senseless death. Tell him if Roma reaches her arm to punish there will be empty huts in his land, black fields and wailing. Tell him this.'

He answered briefly, and spat.

She said, 'He does not see these men.'

'I don't see the sun. But it surely will rise.'

She spoke, quietly. When she had finished he began to laugh. The noise spread. I stood with my hands clenched till it had finished. He leaned forward then, talking earnestly.

She turned to me unwillingly. 'He says he sees the empty huts, but they are yours. The great huts with the golden roofs. He says there is a prophecy. The fox will watch from your windows, badgers roll in your courts.' She swallowed. 'He says you are under the protection of a Scotic princess, and will therefore not be harmed.'

'Tell him,' I flared, 'I'm under the protection of Honorius Augustus and the strength of Rome. Tell him he's a barbarian; and ten times a bloody fool!'

'If I say that,' she said calmly, 'he'll kill you.' She spoke again, liquidly. Another shout of laughter; and he bowed to me, exaggeratedly, before wheeling his horse. Something showered at my feet; a handful of golden coins. I stepped forward, shaking with rage; and an arrow sang past my face, embedded itself quivering in the door-frame. I froze; next moment the riders had bunched together and turned, were pouring away across the grass. For a while their yells echoed back to the headland; then the night had swallowed them, and there was quiet.

I ran into the building. I said, 'Hurry, Crearwy. Hurry.'

'What are you doing?'

'What do you think?'

'It's all right,' she said. 'They've gone.'

I straightened slowly, a saddlebag in my hands. 'What? What did you say?'

She said, 'They've gone.'

'And what do you think I'm going to do? Lay here while the bloody Province goes up in flames?'

She shouted back. 'There's no need to take it out on me!' She began angrily slapping things together. She said, 'I'm sorry your dignity was ruffled. You could have been killed.'

'I'd have taken him with me.'

'Oh, don't be so silly!'

'Will you get the horses?'

She sat back on her heels. 'Your trouble is,' she said, 'you can't stand being laughed at.'

'It's hardly a laughing matter. In any case, you needn't have joined in.'

'I didn't!'

'You damned well did!'

'It was laugh or be killed,' she said. 'He wasn't a patient man.'

'Neither am I!'

'That,' she said bitingly, 'I'd noticed. . . .'

I went for the horses myself. By the time they were harnessed she had finished making up the packs. I mounted, took the reins of the baggage animal. She said, 'If it's not a military secret, where are we going?'

'Portus Adurni. I'm rousing the garrison.'

'I can't!'

'You will.'

'*I will not!*'

She stared at me tearfully, breasts heaving.

'Crearwy,' I said, 'what you don't seem to realise is that I have responsibilities in this Province. Don't you understand? These were barbarians. There were nearly a hundred with him. How many more are on the rampage God only knows.'

She tossed her head. She said, 'I'm not keeping you from your duties. Go and get your soldiers. Get as many as you like. I shall be all right.'

'Please don't be difficult.'

'Hah! Hark who's talking!'

I said, 'You know I can't leave you.'

'I'm safer with them than you!'

My temper snapped completely. 'Who's side are you on, anyway?'

'Not yours, while you go on like this!'

'For God's sake . . .'

She yanked her horse round, swung into the saddle. 'What you seem to forget,' she said, 'is that they're my people. If you're going to make a clean sweep of things you'd better get rid of me as well.'

I flared back. 'And what you seem to forget is that all you own, all your husband owns, you owe to Rome, not a bunch of bloody roughnecks in Dalriada, living in mud huts and lifting each other's cattle. . . .'

That was as far as I got. She beat at the horse, furiously, with her fist, broke into a gallop. I followed her, cursing myself and all barbarians, pacing her shadow across the moon-whitened grass.

The sky paled by degrees. We trotted silently, side by side. Her lips were white, her face set like stone. She only spoke once more. She said, 'He would have protected me. Roughneck or not.'

'Who?'

'Ossa. The King.'

I was still too angry to answer.

The dawn was grey in the sky before she reined. She said, 'I'm sorry. I can't go any farther. I must rest.'

The rage had gone. In its place my head throbbed sullenly, as if I'd had a night on the wine. I stared round me. I said, 'There's nowhere here to stop.'

We walked the horses again. Eventually we came to a farm.

It wasn't much of a place. There was a stockade, its gate hanging ajar; a couple of circular thatched huts, sunk half into the ground. I rode to the larger, leaned to beat at the low door. A dog set up a yapping inside. I hammered again, with the pommel of my sword.

The door creaked open. The woman who stared up looked frightened and old. She held a blanket clutched across her throat. I said, 'Do you understand Latin? We are tired. We need rest, and some food.'

She nodded, silently, waving for me to enter. I swung stiffly from the horse, stooped under the lintel. The inside of the place was primitive. I saw a beaten-earth floor, a table and oven built of slabs of stone. To one side a fire was burning; over it hung a blackened iron pot.

Crearwy followed me. I picked up an earthenware cup. It didn't look too filthy. The woman was kicking at a bundle of rags in the corner. 'Get up, you old slug,' she said. 'Here's a great Roman soldier, and his lady.'

'Peace, mother,' I said. 'Let him sleep. We can help ourselves.' I skimmed the coating of grease from the stew, handed the cup to Crearwy.

She squatted by the fire, pushing back her hood. 'Sergius,' she said, 'you are a strange man. When shall I understand you?'

I stared at her. I said, 'Perhaps when I understand myself.'

We rode on. At Sorviodunum there was a way station. I changed horses. The sun was high now and we still had a great distance to go. At Cunetio I requisitioned more remounts. After that at least the way was firmer. We were on the Corinium road.

Crearwy had relapsed into silence and I didn't feel like talking. Twice in the night I had seen distant fire in the west. I knew the speed at which barbarians can move. Impossible to guess where these might be, in which direction they would strike. Where would they have beached their ships? The Sabrina, in all probability. They wouldn't risk the Sea of Vectis again. Not so soon. I cursed, urging my mount forward. The road seemed endless.

It was late in the day before we reached Corinium, and Crearwy was reeling with fatigue. I reined, gripping her arm. Horsemen were riding to meet us from the city.

It was Riconus. He glanced at her keenly but made no comment. She sat her horse a little way apart while we talked. What he had to tell me set me cursing again.

The garrison of the city was alerted and standing to arms. We clattered in through the gates. I made for my office, helped her dismount. I said, 'You must rest.'

'I can't. The children.'

'You're not going to Censorina. It's dangerous. You can stop with the Praeses and his wife.'

'What are you talking about?'

'I can't tell you now. I'll send the children to you. Riconus will fetch them. You musn't leave Corinium, do you hear?'

She said, 'I hear.'

I left her, hurried to the basilica. I needed men. I drafted urgent orders for the Germans at Glevum, wrote a hasty despatch to Hnaufridus and saw it on its way. Whether it arrived or not I never knew. By that time messages had started to come in from the south. Aquae Sulis and Abonae were under attack and requesting help. I tried to get some idea of the numbers involved. It was impossible. To a frightened town guard, ten barbarians make the same impression as a hundred. I snatched a bite to eat, ran back to my office. The lamps were burning; but the place was deserted. I turned as Petronius came through the door. He backed off at sight of me, looking alarmed.

'Where's the Domina?'

He said, 'She's gone.'

'*What?*'

He spread his hands. 'What could I do?'

'Gone where, man? Where?'

'The villa. How could I stop her?'

I said, 'Get out. *Get out.* . . .' I sat, rubbed my face and tried to think. I was still sitting when Riconus hurried in. Another messenger had arrived. Scoti in numbers had been seen on the Calleva road. Farmhouses were burning, and Calleva itself besieged.

It had been three hours since I sent to Glevum, ample time for fast-moving troops to reach me. I could wait no longer; at this rate it would soon be dawn. If Crearwy was at Censorina there was an end of it. I had done what I could. I said tiredly, 'Get your men assembled, Riconus. We shall ride.'

The camp, if camp it could be called, sprawled across open

land ten miles east of Corinium. There was light in the sky by the time we reached it; enough light to show its extent. At my side, the Celt sucked his breath. It was big, far bigger than either of us had believed. How many men it housed I had no idea. To the right, a hundred yards or so away, stretched a line of waggons; farm trucks and carts, an army baggage vehicle, an ancient, dilapidated carruca. Beyond them were the tents, sprawled anyhow across the grass. They were flimsily built of the first material that had come to hand: rugs and blankets slung on poles, pieces of board and sacking. Smoke curled up from cooking fires. Everywhere men stood about in sullen, ragged groups. As they became aware of us, they turned to stare. Nobody spoke or moved. There was no sound at all.

I glanced behind me. The Palatini had drawn up in a tight double file. Their faces were impassive. I said, 'Move forward. At the walk.'

We passed between the first of the tents. Everywhere the faces watched in silence. A child bawled, and was quieted. Riconus eased his sword in its scabbard. He said between his teeth, 'Where have they all come from?'

I said, 'I don't know.'

Maybe I didn't. But I could guess. They came from Deva, from the unheated barracks, the Praetorium where every morning they bundled up and buried the corpses of the dead. They came from Lindum and Eburacum, from Manucium, from the shattered townships of the Wall. They were the homeless and the hopeless, the hungry, the frightened, the bewildered. They were soldiers who'd seen their armour rot and their pay diminish and their horses die, slaves who'd known no life but the life that was gone, burned with their masters' homes. They were men who had seen their families destroyed, youths who had lost their parents; they were children and women, farmers, peasants, beggars, sailors, whores. There was no understanding in the faces we passed; no compassion, no humanity. There was fear, and incomprehension, and a dull, smouldering hate.

They were headless, brainless and strong. They lacked intelligence, they lacked direction, they lacked a leader. But a leader, it seemed, had been found. And three days had been enough; enough for the ripples to run out, the murmur of uncertainty and fear. Enough for the word to form.

Bacaudae . . .

There was a Praetorium, of sorts. Above it hung a flag. It was ragged and crudely stitched, but already it had drawn men from half across the Province. I strained to make out the device it bore. As we drew nearer I recognised the mark of the Christos.

I reined my horse. A wait; and the flap of the tent was raised. The man who stood there was bearded and dishevelled. A naked sword was stuck loosely through his belt, next to it a dagger. He gestured, silently. I dismounted, with Riconus. I had a sense of unreality. I ducked under the leather. The inside of the pavilion was dark. At its far end a man sat at a table. A single lamp burned; it showed a wine flagon, a shallow basket of bread. It also showed his face; the stringy hair, pale skin and eyes. The eyes were fixed unwaveringly on my face.

His Latin had improved, but the harsh voice was the same. 'Enter,' he said, 'and be welcome. Sit down, Roman, and eat. Once I shared your table. Now I think you'll share mine.'

I walked forward, one hand on the pommel of my sword, and stared down at him. 'You always were a fool, Ulfilas,' I said. 'And you don't seem to have changed with time.'

Chapter nineteen

He grinned, with his lips. 'It's you who are stupid, Roman,' he said. He nodded at the tent flap. 'Where are your eyes? You've seen my people. One word, and you are dead. You and all your soldiers.'

I shook my head. I said, 'Don't think of it. You're dealing with Palatini, not a rabble of slaves with sticks.'

The smile stayed fixed round his mouth. He said, 'And you're dealing with two thousand men.'

'Ulfilas,' I said, 'I didn't come here to bandy words with you. So listen, and listen carefully, because I shall only tell you once. I don't know what your intention was in gathering these folk together, but you're leading them, and yourself, to disaster. The country hereabout won't support large numbers of refugees. There's no work for you, no shelter, no food. Already your people, as you call them, are probably suffering from illness. Soon you'll have disease; from bad food, bad water, bad sanitation. You'll not seem much of a saviour, in their eyes or anybody else's, once that starts to happen. If you've got any sense left you'll know I'm speaking the truth. So this is what you'll do. You'll use your influence to disperse this mob now, while you have the chance. The towns in my care can absorb a limited number of them; the women and children at least, and husbands of families. The rest must go. Since I don't choose to see people starve on my doorstep I'll send you grain supplies for forty-eight hours. After that time you'll get nothing. While if you persist in this absurdity I shall undoubtedly hang you.'

He said, 'We shall not starve.' He rose, came limping round

the table. One leg, I saw, was withered; it had been smashed, and badly reset. He saw the direction of my glance, and sneered. 'Roma gave me this,' he said. 'I left the mines a free man. Yet she followed. I was left for dead. But the strength of the Lord was with me. By his hand, I was delivered from my oppressors.' His eyes blazed, hooded themselves. 'By his hand,' he said, 'these people will be fed. By his hand they will clothe themselves in soft garments, by his hand they will be sheltered. For it is written, the Beast that sat on seven hills is dead. The lowly inherit the earth.'

I said, 'They'll make a saint of you yet.'

For answer, he spat accurately at my face.

For all his bulk, Riconus could move like a cat. One moment Ulfilas was standing glaring at me, the next instant he sprawled across the table, felled by a blow from the Celt's huge fist. The bearded guard leaped forward, grabbing at his belt; but he was far too late. My sword-tip was already pressing into his neck.

'Unfasten,' I said. 'Let your weapons fall. Slowly.'

Riconus grunted disgustedly. One brawny arm held the Goth in a vicelike grip; the tip of his dagger was laid against his victim's throat. He said, 'I wonder if his God would accept him minus his tongue and balls?'

I shook my head. 'You won't frighten him,' I said. 'He can only be killed.'

Ulfilas grinned lopsidedly. 'Remember this, Alcimer,' he said. 'Never trust a Roman.' He turned his head, with difficulty. 'Now,' he said, 'what do you intend? Kill me, and you'll never leave here alive.'

I knew, with a sick certainty, that it was true. I'd allowed tiredness and anger to cloud my judgement; I should never have entered the camp. A determined charge might take a few of the Palatini clear, but I would never drag him with me. Neither dare I leave him. I said greyly, 'I think that's a chance we'll have to take.'

The dagger rose. He said quickly, 'Wait.'

Riconus hesitated.

He licked his mouth. 'It isn't too late,' he said. 'I could use your men. Join with me. The Province is rich. Rome's priests are false, the Emperor is a child. Here, under God, we could found our kingdom. All men would be equal in the sight of the Lord.'

I said, 'You're quite definitely mad.'

The dagger shook again. He said desperately, 'No. There's another way. Kill me and you'll regret it.'

I said, 'Since when did you fear death?'

He writhed, wincing. Riconus tightened his grip. He said, 'I live for my people. We mean you no ill. I'm no use to them dead.'

I sheathed my sword, walked across and sat on the edge of the table. I picked the wine up, sniffed it and set it down. I said, 'I can't afford to take your word, Ulfilas.'

There was something like eagerness in his face. He said, 'I wouldn't ask you to. I'm prepared to give an oath. One I can't break.'

I stared at him, wondering how much he knew. I could contain him, with the men at my disposal; but the cost to Prima, already heavily infiltrated by Scoti, would be severe. I needed time; two days, three at the most. If I couldn't win them, they would have to be bought.

He said, 'Let Alcimer fetch the priest.'

Another wait; and I shrugged. I said, 'As you will.' Whatever farce he hoped to perform, it seemed best to let him see it through.

The bearded man turned with a final glare, and stalked from the tent. I heard him bellowing outside. He came back shoving a Bishop in front of him; a frightened-looking little man, fingering the massive cross that hung about his neck.

Ulfilas said, 'Give it to me. Let me touch it.'

I nodded to Riconus. He relaxed his grip, unwillingly. The Goth took the crucifix, ran his thin fingers across it, pressed it to his mouth. 'By this I swear,' he said. 'The symbol of our Lord. No harm shall come to your territory or your people, while Britannia holds faith with Rome. As she is true to the Empire, so I am true to you. Or I renounce my God, willing my spirit to the flames of Hell.'

The sun was well up by the time we rode away. A mile from the camp Riconus turned to me. He said, 'I'd sooner trust an adder not to sting than take his word.'

I said, 'He gave it freely. What game he's playing I don't know, but he can't recant.'

'Why not?'

I said, 'Because he's mad. And I can't be in a dozen places at once.'

He said curtly, 'No. Of course not.'

'Riconus,' I said. 'What else could I do?'

He stared at me under his heavy brows. He didn't speak again.

At least the Corinium garrison was doing its job. Traffic was moving normally into the town, under the watchful eyes of sentries. I hurried to my office. In the doorway I collided with no less a personage than the Praeses of Britannia Prima himself. A mule train had been attacked, at Spinis on the Calleva road, its drivers massacred and its waggons left to blaze. Cunetio, slightly to the west, reported sightings of barbarians. Troops from the little town had attempted to engage. No further news had come through.

I spread a map out. I said, 'They'll run rings round them. Or cut them apart. Riconus, can we find a messenger?'

'I'll try.'

I said, 'When I want them to move, I'll tell them.' I stared at the map. It seemed we were faced with two main concentrations of raiders; a western party, still somewhere in the vicinity of Aquae Sulis, and an eastern group straddling the Calleva road. These last, presumably, under the command of Ossa himself. I swore, and banged the table. It was guesswork, guesswork all the way. Like fighting shadows.

Riconus was back. He said, 'There's nobody.'

'Then send to the posting station.'

'I did.'

I turned. 'Lord Praeses, I shall have to call on your staff.'

He drew himself up. He said, 'Except in grave emergency...'

'This is a grave emergency. Petronius, get over there. The first man you see rides to Cunetio. Riconus, we're going south.'

'South?'

'Yes,' I said. 'Look here. If this party are fat they'll be moving west. We'll intercept along the line Corinium, Aquae Sulis.' I frowned. 'Thirty miles. That's twelve hours' marching, with breaks. Call it fifteen. We shan't be there before midnight.'

He said, 'What about Calleva?'

'Calleva has troops of its own. Also it's nearer Count Hnaufridus than me. He'll have to relieve.'

He said, 'Who will you be taking?'

'The town garrison.'

The Praeses said, 'I must protest . . .'

'These are my decisions, sir.'

'The town is my responsibility!'

I said, 'And the safety of the Province is mine.'

'I shall report your actions to the Praefect of the Gauls!'

'Report them to the Devil.'

He said furiously, 'They won't march.'

'Who won't march?'

'The garrison.'

'I'll hang the first man to refuse.'

'I shall speak to them myself! I shall forbid it!'

'Riconus,' I said. 'Escort this gentleman to his house. See he stays there.'

'This is preposterous! I refuse!'

'Then be silent. Who's there?'

A scuffling at the door. Petronius said, 'It's the Germans, sir.'

'What Germans?'

'From Glevum.'

'Thank God for small mercies. Just what took them so long?'

'Messenger, sir. Begging your pardon. Wouldn't ride till daylight.'

'Have you got his name?'

'Yessir.'

'Remind me to have him flogged. Riconus, this gives you a hundred and fifty men. I want you to move south-west, to the coast. They'll have left a rearguard. Engage it, and destroy their ships.'

'How do you know the ships are in the Sabrina?'

I said, 'I don't.' The Praeses snorted. I glared at him. 'A spare mount for every man. If you can't find enough horses, commandeer them.' The Praeses opened his mouth to argue. I said, 'You can bill the Army later.'

I spent the rest of the day on the march, chafing at the slowness of the infantry. Before they'd gone half a dozen miles they were complaining of blisters and fatigue. The delays did nothing to improve my temper. I hadn't seen fit to explain my thinking to the Praeses; but I was taking a considerable risk. I'd left not much more than a token force holding Corinium; everything I'd learned of seaborne barbarians convinced me my presence in strength between him and his ships would bring Ossa scurrying west. He certainly wouldn't turn aside to besiege another fortified town, however tempting it might look as a prize. The real gamble, of course, was whether the ships were in the Sabrina at all. But they must be; commonsense told me that.

The cavalry rode with me for a time before turning across country. I didn't want my forces too widely separated. The manœuvre was at least partly successful. A strong concentration of the enemy—reports varied between two hundred and two hundred and fifty—took alarm at our approach, bolted west into the arms of Riconus and the Germans. A confused running battle followed. Many of the Scoti fell. The rest bundled into their ships, headed out into the Sabrina. There nature took a hand in finishing what I had started. At certain seasons of the year violent tides race up the estuary from the sea. The boats, caught by the full strength of one such bore, were swept into confusion. Several capsized; the crews of others, struggling ashore, fell easy prey to bands of Silurians patrolling the farther bank. Riconus, pushing on, ran into a well-entrenched position. Stubborn fighting ensued. In the end the Scoti retreated. A dozen vessels were burned on the shore, several more disabled. In the meantime my scouts reported barbarians moving west on the road from Verlucio, pursued by garrison troops. Tired as they were, I faced the men about. The Scoti, moving in some disorder, collided with the tail of the column and recoiled. They fought savagely, neither asking nor giving quarter; but trapped as they were between two considerable forces their position was hopeless. A few escaped into the gathering dusk; the rest fell where they stood. We counted fifty dead, recovered six waggons loaded with booty. These I took with me to Aquae Sulis to wait redistribution to their proper owners. My losses were twenty men.

I spent the night in the town. Towards dawn Riconus rode in. The estuary was clear now of enemy vessels but the situation was still confused. Fires had been visible, reflecting from the clouds. It looked as if either Veta or Isca of the Silures was in trouble.

I'd had about four hours' sleep in the last two days. My head was swimming with tiredness but it was impossible to rest. What might be happening at Calleva I had no means of knowing, but I'd failed in my primary object: the main force under Ossa was still presumably in the east. I paraded my combined strength at first light, force-marched for Verlucio. At mid-morning a messenger reached us, riding a jaded horse. The enemy had been seen moving insolently enough along the main road towards Durocornovium. Germans from the Saxon Command had reached Calleva, but little more was known.

There was no help for it. I detached the cavalry, with instructions to hold the town at all costs till the main column arrived. Once more it was nightfall before we marched in. I'd been ten hours on the road; ten hours of threatening, bullying, cajoling, haunted all the time by a persistent fear. If the Scoti avoided battle, nothing stood between them and the capital of Prima. If they once reached the town the Praeses, assuming he survived, would have my head. The thought stirred another, equally unpleasant. I'd offered to put a senior official under house arrest; whatever happened, there'd be the Devil to pay. Well, be it so.

Riconus reported no sign of the enemy. His scouts had likewise failed to make contact. The Scoti had vanished like ghosts; and my men would march no more. For the moment, we were through.

The place was little more than a village, sheltering a huddle of frightened inhabitants. Every vacant house, and many of those still occupied, had been pressed into service as emergency billets. Sentries had been posted. I strengthened the guard. My object was twofold. I couldn't trust my infantry. There had been grumbling in the ranks all day; I had no intention of seeing my strength sapped by wholesale desertions.

I struggled out of my harness and uniform. My eyes felt sore, my throat was harsh with shouting. I drank a little wine; but I was too tired to eat. I spread a mattress alongside the Celts. It seemed I was unconscious as soon as my head touched the rolled cloak that served as a pillow. I fell into a dreamless pit of sleep.

Somebody was shaking my shoulder. I staggered back to awareness. I thought I'd been asleep for hours; but round me the lamps still burned, tired men settled and grunted. I stared up blearily. I heard my mouth say, 'What the Hell's wrong now?'

The first words had me on my feet. Before the man had finished I was shouting for new horses, and for my armour to be brought. I ran into the street like a man possessed.

The Celts rode with me. We struck due north, by farm tracks and lanes. The night was cool and overcast, the air sharp. It cleared my head a little. We crossed the Verulam road, turned north again. I kept up a killing pace; but fast as I moved, the terror reached before. A mile from the road I saw the glow ahead against the clouds. I'd seen it too often in the past few days not to recognise it for what it was.

The place had been a palisaded village. The flames roared fiercely, a great pyre lighting the sky. I reined a hundred yards from the gates. A hut collapsed, with a crackle and roar of sparks. The heat beat against my face.

There was an old man. He cackled when they dragged him to me, mumbling that the Roman wolf was dead. I took him by the hair, pressed my swordpoint into his throat. Blood coursed instantly across his chest. 'Speak quickly,' I said, 'or you might feel her teeth.'

Old as he was, the love of life was still strong in him. He coughed and gargled, rolling his eyes, scrabbling at the front of his tunic. I grabbed the packet he held out, flung him away. I turned the thing over. The seals I knew; the other mark, the brown, irregular splotch, was blood.

The flames gave light to read.

To the Praefect C. Sergius Paullus, at the garrison of Corinium.

The shield has been raised; I am no longer controller of my fate. Britannia wills her own destruction. God grant that I can stay her hand . . .

I am dispatching five hundred men, under the Tribune Libius Naso. He is empowered to act for me in all things. From you I shall of course require undertakings of loyalty. The Province must stand undivided. Your oath to Naso will be as your oath to me . . .

I have sent letters to Stilicho and the Emperor Honorius, claiming the recognition due to me. God knows, and you perhaps may guess, how far this was from my will.

M. TAMMONIUS VITALIS,
AUGUSTUS

The words seemed to swim in front of my eyes. I crumpled the thing, flung it away. It was clear enough, now, how I had been duped. A phrase came back to me. *As Britannia is true to the Empire, so I am true to you.* Twice I had jested with him. Now he had jested with me.

I turned to the Palatini. '*Ride*,' I screamed at them. '*Ride* ...'

Everywhere in his track the night had blossomed fire. The flames lit hill-tops and empty fields, silent, brilliant, impossibly red. Columns of smoke trailed slowly, leaning with the wind. The air was acrid with the smell of burning.

The glow from the little valley brightened the land for miles. I rounded the last spur, saw the villa below me. If it's possible for a soul to die, I think mine died then. Censorina was in flames.

They had overturned waggons, for a breastwork in front of the gate. Hurdles they'd used and rope, grainsacks and beams and hides. They tore the spiked holdfasts from the windows, scattered them for caltrops. They fought with javelins and and swords, pitchforks and axes and spades, cleavers from the kitchens, the carthouse chains. They fought, in the end, with fists and teeth and nails. And they were overwhelmed.

The nearest of the carts had been dragged aside. I set my horse at the gap. The creature screamed and reared.

He was standing by the inner gate. Behind him were flames. He limped forward. He leaned his weight, heavily, on the broken shaft of a spear. How he moved at all I shall never know, for one ankle had been hacked to a wet stick. He said, 'I waited for you.'

He stared up blindly. His face was striped with blood. 'We held them here,' he said. 'The second attack was from the south. They came over the roofs. You killed us. You killed us all.' He staggered. 'I left Pelgea,' he said. 'Will you take me back to her?' His hand gripped the doorpost, convulsively. The wood crumpled under his fingers, disclosed a living heart of fire. Sparks whirled in a column as he fell. I heard my own voice shout, '*Valerius!*'

I leaped the horse across him. Heat and brilliance burst on me. The lawns, trampled now and bloody, were lit brighter than day. I fell from the saddle and ran, shielding my face and yelling. '*Crearwy* ... ! *Crearwy* ... !'

I found Pelgea. She lay face down across the triclinium steps. She was naked to the waist. The blood she had voided had spread round her in a pool. A dagger had been driven into her neck at the base of the skull. Near her, contorted, was the body of a man, hands gripped across the face. Between the fingers stood the shaft of a hare-spear. The prongs had been driven through the eyes, into the brain.

I ran again. It seemed the pavement flowed beneath me of its own volition.

'CREARWY...!'

She was seated by the nymphaeum. Beside her, a little distance away, lay two small bundles of sodden cloth. She looked up at me. Her hair was unbound, one shoulder bare. A bruise showed on the flesh.

'He left me for you,' she said. 'It was his will. He wanted you to know his name. Ulfilas, the Goth.' She turned, vaguely, to stare. 'Nessa distrusted me,' she said, 'because her arm was gone. I called her, but she wouldn't come. Melinda was thirsty. She came to the pool; but the blood came from her nose, and spoiled her drink.'

I reached, slowly, to raise her, and for the first time she twitched away. 'Now I should curse you,' she said, 'after the manner of my people. By the Hand, and the Branch, and the Hounds of Bran.' She flung herself down then, silently, lay cramming dirt and ashes at her mouth.

There were hands on my arm. It was Riconus. He led me to the steps, walked back to where the children lay. I put my face in my hands. Through my fingers I could still see the glare of the flames. Their roaring seemed to fill my brain.

The dawn came slow and grey. Light stole across the waste of ash. What flames still burned were paler now. Smoke drifted from the ruins; through it gable ends showed stark, like bones. The lead had run from the bath-house tanks, lay congealed in puddles and streams.

There was a burial. The Celts dug the shallow pit, in the lawn by the smashed roses. Riconus lowered the children's bodies into the soil, making an offering of coins and birds. Later he came and squatted by me, speaking to the grass at his feet. He said, 'The reinforcements are here.'

'What reinforcements?'

'The men I summoned for you.'

I rose slowly, and followed him.

Some sat their horses, waiting. Others squatted round fires and hastily pitched tents. They were from Cunetio, Durocornovium, Verlucio. I rubbed my mouth. The ash-taste was thick in my throat. I said, 'Give me some wine.'

A skin was passed to me. When I had drunk I said, 'There is

an Emperor in the north. Which of these men would least gladly follow him to Gaul?'

He nodded at a group of some twenty or thirty, sitting silently by their horses. He said, 'They are foederati, from beyond the Wall.'

I said, 'Then they know her father for a King. Where is she?'

They had saddled a horse for her, hung a cloak round her shoulders. She sat the creature rigidly, staring in front of her. Her face was marble.

'Riconus,' I said. 'Give them some gold. It will be repaid you.'

It was done, silently. Then I walked forward. I said, 'The King will also reward you. Take the Domina to her home.'

Someone, I saw, reached to grasp her reins. Once she put the hair back from her face, but she didn't turn. The column moved away, down the long hill of grass. As it receded there came from it the squeal of pipes. No dirge they played but a marching tune, wild and defiant and gay. The sound reached me, borne on the wind, long after the last of the riders had dipped into the fold of ground and was out of sight. I stayed where I was, staring. In time I saw, or thought I saw, climbing the opposite slope, a moving patch of greater darkness. The wind blew again, gusting fitfully. It seemed it carried with it a ghost-thin shred of melody. Then that too was gone.

I thought the shadows were in my eyes. But it was the night.

The hooves of the horses rang dully, now close, now far-off in my brain. Grass jerked by half-seen and bushes, the metalling of the road. Behind me poured the column. As it moved so its numbers increased; for my messengers had gone out to every town and hamlet within a day's ride, stripping them of men.

There were towers and walls. I understood that I had reached Corinium. The army rested overnight. I sat in what had been my office. The room now was haunted. I dozed and woke, each time with a lurch like the falling nightmare; and somehow between the eyes and hands and hair, the temple and flowers and the noise of flames, the dawn came.

The Wall contingents had arrived. I turned them west along the Glevum road. Round me the world spun and hummed; but my brain and will remained clear. Loyalty and honour, dignity, right and wrong, all shadow-things that had bedevilled me, seemed swept away. I was lightened by their going. My life, I saw, had acquired a purpose. The path, the time remaining, stretched straight and terrible. My enemy fled before me; my sole charge was to follow, if necessary to the edge of the earth.

'Riconus,' I said. 'Add to your requisitions, a waggon of salt.'

'Salt,' he said. 'What for?'

'This Province will remember Rome. And Riconus . . .'

'Sir?'

'Where is the Standard you found at Deva?'

He said, 'In my pack.'

'Then raise it. One is as good as the next.'

In the void in which I moved, strange fancies came to me. My army, I realised, was a sentient thing. The scouts I flung out to either side of the road might be likened to the antennae of an insect, touching and delicately probing. The Palatini represented the head and brain. Behind, inching and glittering, moved the great scaled body. My will sustained its progress; its weight bore down on me, carried on my shoulders alone. The din, the tramp of feet and shouting, endless crash and rattle of waggons, added to the burden. The faces mouthed, but it was difficult to bring my brain to bear.

There were riders; Germans, surely, and farmers to boot. But each man carried the seax-knife, the sacred dagger from which their nation took its name; and each rested a heavy battleaxe across the neck of his horse. They checked at sight of the column, and their leader rode forward alone. 'Once I listened to you, Roman,' he said. 'I and all my people. Now my sons and wife are with the Shades. There is death between us.'

I dragged my mind from distance. The Celts, I saw, had swung outward in a threatening half-moon. A sword rasped clear of its sheath. I said, 'How are you called?'

He stared round him, slowly. The lank hair blew across his face. He said, 'Gundebad, son of Gontran.'

'Death I can bring you quickly, Gundebad, with either hand,' I said. 'Death, or vengeance. Which is it to be? The choice is yours.'

He waited. Finally he said, 'Vengeance.'

'Then by all the Gods you shall have it. Get your men into line. . . .'

The weather broke, in flooding rain. The water streamed across the rutted surface of the road, gleaming on the high shoulders to either side. I lived with the stink of wet cloaks and leather, the creak and rumble of wheels. The column, butting into the deluge, seemed haloed with dull silver. Prisoners were brought to me, stragglers from the Bacaudae. I gave them to the Germans. Some they bound, and sat on sharpened stakes. They were left behind, dark things that moved and begged. Later the memories returned once more to trouble me. I drove them away with wine. I needed Ulfilas now, the sight of him, with the urgency of a lover. At night I needed him, in my tent when the wine had gone.

The column passed through Glevum. The western gate had been fortified with a breastwork of dressed stone. I ordered it torn aside. Camp was pitched where a tributary of the Sabrina flowed from the hills, spreading into a series of shallow, reedy pools that enclosed a narrow spit of firmer ground. The tents—good military tents, of stitched leather—went up anyhow, guy lines crossing and recrossing. The baggage waggons I had drawn into a line across the neck of the spit; on its other flanks the camp was protected by the marsh, through which no attacking force could splash without giving itself away. Sentries were posted; and in time the last password was exchanged, the last man fell over a half-buried tent peg and the camp settled to an uneasy sleep.

In the nightmare in which I lived, days and nights had blended into a grey uniformity. I wondered, with a part of my spinning brain, how long it had been since I had eaten or removed my clothes. Yesterday, or the day before, somebody had offered me a plate of gruel that I refused; since then they had let me be.

Spatters of rain drove like slingshots against the tent. Above, the sky was clearing. Stars glinted in the gaps between the clouds, but the air roared like a forest.

The voice was in the wind, sweet and wild. I started up. Hands plucked the tent fastenings. I shouted, reeling across the little lighted space. The lashing of the tent flap, swollen with damp, defied my fingers. I plucked and tore, desperately. The voice keened again; there were other hands, yanking the stiff fabric. I panted, fighting with the leather and cords.

The winds raged, in and across the tent. The lamp flame

surged, was extinguished. I fell forward in darkness, landed on wet grass; and she was gone, slipped through my fingers like a wraith, back up into the bawling sky. I lay tangled among ropes, hearing her echo and my own hard sobs.

There were lanterns, and a torch. The torch streamed in the wind, a bright beard of flame. More hands gripped, trying to raise me. I called, 'Valerius . . .' and heard Riconus swear.

They lit the lamp again, flung a cover across my legs, put the wine out of my reach. They stood to till dawn, turn and turn about, before the tent where I fought, I suppose, with Devils.

At first light the pickets came in, stood swearing and shivering in their damp cloaks, chafing hands and arms, kicking smouldering logs into a blaze. Clattering rose from where women—two contingents from the north had arrived with their entire families—stirred vats of soup and porridge, ladled the scalding mess into the pannikins held out to them. Fatigue parties scurried round the horse lines and crudely dug latrines; others, armed against surprise attack and well muffled in the chill air, set out to forage brushwood for the fires.

Dawn brought a contingent of black-robed Priests of Nodens. They had news for me; the enemy, sluggishly alarmed, had turned once more for the north. Weighed down by booty and harried by the half-wild tribesmen of the hills, they were making slow progress.

The tents were struck, with more haste than efficiency, the mule teams harnessed to the carts. The column formed into its line of march, with much shouting and bullying of laggards; and the Palatini moved away, stepping in single file along a rough track that wound through scrub and woodland to the north. They rode carefully, on the watch for ambushes; ahead the scouts were active again, beating the ground to either side of the path.

By midday the leading cavalry were clear of the belt of woodland. The sun, breaking through the clouds, struck gleams from lance-tips and the bosses of shields. No opposition had been encountered, but signs of the enemy were more frequent now. A cluster of huts, set prominently on a wooded spur, smoked sullenly; farther on were the bodies of a dozen women and men. They lay tumbled where they had fallen, the trampled grass around them stained with red. The cavalry passed with an indifferent glance or two; professionals all, death held no fascination.

Time after time the column, pushing on dourly, was forced to cross streams. This whole area was laced with watercourses, tributaries of the Sabrina; the brooks raced and gurgled, swollen by a day and night of rain. Time after time the waggons bogged at fords; the limitanei, sweating now under the warm September sun, cursed as they heaved at the axles, waist-deep in swirling water. One plaustrum overturned, spilling its cargo with a great splashing. I ordered its trace animals cut free; they could be put to better use elsewhere.

The sun was sinking below the western rim of hills before Riconus wheeled, thrusting the Standard into the turf to mark the resting place of the praetorium.

I rode forward alone, leaving the plateau selected for the night's halt bustling with activity. Ahead, the ground swept up gently to a wooded ridge. I climbed it, sitting my horse slackly, letting the animal pick its own way to the crest.

On the skyline the trees stood dark against the sunset. I rode between them, passed from gathering dusk back into light. Before me stretched a shallow valley, five miles or more across and carpeted with woods. To my left, the levelling sun struck between two low humped hills; the eastern slope, bright already with the colourings of autumn, seemed to flame in the pouring orange light. The wind had dropped, through the day; now no breath of air was stirring. The valley lay still, and totally quiet. Nothing moved, either human or animal; the leaves of the nearer trees hung like translucent golden coins.

There was a smell of burning; faint yet pervasive, as if the land was indeed ablaze, smouldering with all the fires of autumn.

Beyond the far rim of the valley, barely discernible against the distant outlines of hills, blue-grey threads wavered upwards. There were many fires; over the ridge, hidden beyond a further belt of woods, an army had come to rest.

I watched for a long time, then wheeled my horse and rode at the gallop back down the slope to the camp.

The woods lay silent, velvet-black in the night. Badgers snuffed and grunted; somewhere an owl called, quavering and high. The column picked its way cautiously; sword-hilts clinked, once a horse shied and snorted at some looming shape above the path. Round the trunks of the trees mist coiled in thick, slow-moving veils; more than one of the Palatini crossed himself as he urged his horse into the shadows. The infantry followed, stepping close behind the trailed lances. For a while, as the army breasted the southern ridge, ragged shapes of men showed fleetingly against the sky; then the valley had swallowed them. There was stillness again and quiet.

At the foot of the first slope a shallow brook meandered through the forest. Beside it, a pale shape resolved itself into the figure of a mounted Silurian. He straddled a white-faced pony, pointing silently to the farther path. As the last man splashed through the stream he lifted his head, hands cupped round his mouth. The owl call floated up again into the dark.

The camp sprawled confusedly at the foot of a smooth slope of ground, dotted here and there with clumps of gorse and bramble. The site had been chosen for its seclusion, and for the stream, not much more than a rivulet, that chuckled and splashed a hundred yards away; but the selection was ill-advised. With the dark, vapours rose from the damp ground, chilling the bones of men who already shivered with a variety of fevers and agues. The more energetic of them had rigged tents amongst the baggage carts that littered the slope, but mostly they lay in the open, wrapped in cloaks, tunics, whatever scraps of cloth they'd managed to find, or steal from their neighbours. The rest of the waggons, drawn into a wide half-circle, afforded some measure of protection; between them, shielded from observation by their high wooden sides, half a dozen fires still smouldered. Now and again a burned-through log broke with a snap and crackle, sending up a brief shower of sparks; the sentries, leaning dozing against their spears, jerked up, alarmed, peered into the slowly drifting mist before settling their chins back on their chests. A child wailed fretfully; the sleepers tossed and muttered, wishing it was day.

It was a wearing business, this game of soldiering. At first, infected by a common enthusiam for destruction, they had marched and fought with a will, spitting young and old alike on their new stolen swords, cheering hoarsely as settlement after settlement went up in a blaze of sparks and crashing thatch. They loaded themselves with spoils, drapes and clothing, weapons, brooches, coins, plate; treasures they had rarely if ever seen, let alone expected to own. As word of the destruction spread their numbers swelled; the homeless and hopeless of half the Province began flocking to them. Immediately, problems started to arise. The uplands of the west could not begin to support such numbers; the army was forced to move continually, foraging farther and farther afield for the bare essentials of existence. Faith is no substitute for a full gut; there's little pleasure in slitting a man's throat if you know he's as poor as you are, that his hut won't yield more than a dozen cupfuls of grain. Squabbles broke out among the victorious soldiery; a half-filled wine-skin, or the carcass of a sheep, became killing matters. Also more and more began to suffer from drinking tainted water; and the fainter-hearted started to drift away, in ones and twos at first, then in larger groups. None of them got far; the hillmen, who had been hanging on the flanks of the rabble for days, stalked them through the gorse, silent as shadows. These Silurians knew their business. There was no warning; you seldom caught so much as a glimpse of them. Just a flicker on the hillside, an arrow coming with a dark flash and *thunk*; and another man would fall over kicking, or kneel carefully and start to sick his blood out on the grass.

The mob pushed on, uneasy and disgruntled; for there were rumours, brought in by laggards, of a countryside on the stir. Arguments raged again. The bolder of the outlaws, still flushed by success, were for making a decisive stand; others, their spirits dampened by the rain, urged a retreat to the north. Cautious counsel prevailed, and the march was resumed. Through the blazing day that followed they kept up a gruelling pace. Needlessly, as it turned out; for when camp was finally made, and they dropped tiredly in their tracks, there was no sign of pursuit. But no Roman, not the Dux Britanniarum himself, could hold limitanei to such a chase; once over the Great River the troops would have turned back, satisfied at having dispersed the immediate danger to their homes. The army got to its rest well content.

Except for the cursed dampness of the ground; and the owls that called all night long, eerily, round the camp. As the hours wore on it seemed the birds drew closer, startling the nervous from their sleep. They sat up, huddled in their blankets, grumbling and scratching themselves. Above, the faintest wash of grey light heralded the dawn; the air struck raw and chill.

They rose, kicking their snoring neighbours as they passed, stumbled to the fires to hold their hands out over the wavering warmth.

As the light increased the tall shapes of the waggons ceased to be flat pale silhouettes, took on form and solidity. Men stood yawning, glancing up at the mist. High to the east cloud streaks showed faintly through the pall, touched with pink by the rising sun; the fires, stirred into life, sent up columns of smoke that swirled slowly, mixed with the thinning vapours of the marsh.

By degrees, the outlines of the surrounding land became visible. Then, man after man turned to stare. A silence fell, broken at last only by the clear high piping of a bird.

Along the brow of the eastern slope, shoulder to shoulder and impassive, stretched my cavalry. The sun, a blind bright eye, glared behind them, throwing long shadows forward against the mist. To the rear the infantry massed solidly, spear-points catching and winking back the reddish light. In the van, above the centre of the line, the Labarum hung silhouetted and stark; beside it gleamed the golden boar.

The silence, that had descended so abruptly, was broken with equal suddenness. A confused scurrying began among the opposing force. Men ran shouting, grabbing up weapons; women called shrilly; children bawled. The many voices blended into a sullen murmur, like the distant noise of a sea.

I sat my horse silently. Behind me, more than one of the hastily levied auxiliaries exchanged anxious frowns with his neighbour. Only the Palatini stayed motionless. They knew, too well, the force and crashing shock of a charge of heavy cavalry. That force, properly directed to its target, will burst through nearly anything. They waited, as I waited, for such a target to form.

And now, above the Bacauda line, rose an obscene parody of the Standard on the ridge. It moved and mocked, jabbed on a pike-end, waving slowly in the early light; its fair hair gleamed, hung round the shaft in spikes and clotted tufts. The answering murmur that ran through the German ranks was no less terrible for its quietness. For a barbarian, and a farmer to boot, might not set much store by his woman; little enough time for love, with fields to hoe and sturdy young to raise. None the less it is a notable thing to see her head laugh at you from a stick.

The lines would now no longer be held. I rested my hands on the neck of the horse and turned to the man at my side. 'Riconus,' I said. *'Destroy me those people.'* I turned my mount then, walked it from the line of charge.

In the quiet the champ and fretting of a bit, the scrape of a drawn sword, sounded clearly. Men glared right and left, fidgeting with reins, eyeing the ground between them and the enemy as they plotted their course through the hummocks and bushes of the slope.

Who first raised the cry nobody could afterwards say. Reedy and thin it was, like the calling of a ghost, but growing magically, strengthening as it spread from mouth to mouth along the line. The shout that for a thousand years had borne the arms of Rome to battle, while a village in a marsh rose to be the terror and glory of the world; heard now from the throats of Gentiles and barbarians, swept away finally in the long drumming, the rising thunder-roll of hooves as the lance-tips dipped, the whole mass of horses and men, steel and flesh and bone, surged at the crest, fell like a stone across the intervening slope.

'MARS, VIGILA . . .'

The charge struck the Bacaudae like an unspent wave. They barely served to check it. Everywhere, men reeled back under the shock. The line of horsemen rolled forward, jamming the struggling mass against the waggons. I saw one heavy cart flung bodily on to its side; another, struck by the full weight and press, reared on end, spilling its screaming top-load into the ashes of a fire. Instantly, the screaming redoubled; and smoke rolled up, obscuring the centre of the fight.

The slaughter wouldn't perhaps have been so total had not a chance javelin, one of the very few hurled, brought down Astyr, a stocky, cheerful cousin of Riconus, who was riding on his immediate right. The shaft, flung with the strength of desperation, struck him full in the face, burst out a hand's-breadth through skull and helmet. He plunged headlong, killed instantly, but his shriek still rose to the skies. It was the first loss the Celts had suffered in battle; and it drove them temporarily insane. Placed as they were at the tip of the charge, their speed carried them clear through the mob into the dead ground beyond. No other cavalry could have checked, but Riconus turned them. I heard his long yell, saw the boar Standard surge to one side. The little knot of men flung themselves with appalling fury on to the enemy's flank and rear. The horses, butting chest-deep, tore swaths through the Bacauda ranks; the swords fell and whirled, red-stained by the morning light. The wing of the battle thus assailed rolled back towards the centre, and the whole fight degenerated to a boiling confusion. The Bacaudae, attacked from front and rear and fatally hindered by their own baggage train, began to panic. Waves of men bolted forward, up the hill. They collided, at the run, with the infantry still pouring down.

The smoke from the burning waggons had thickened now; it drifted low across the field, shot with streaks and flashes of flame. Through it, as it were in isolation, I glimpsed details that will always stay engraved on my memory. I saw a woman, already streaming, hold up a baby to ward off a sword-cut. The blade passed through the child unchecked, carried away her cheek and jaw. The next blow, mercifully, finished her. I saw a man, spitted on a lance, held writhing in the embers of a fire till his head burst and blackened, and he died. Nearer me a creature minus his forearms humped his way industriously up the slope. Behind him through the grass dragged pale-red snakes of entrails.

The din, the yelling and shrieking, clash and ring of weapons, reached a peak. The Standards waved and bobbed, now glinting, now seen as shadows against the smoke. Then, almost it seemed without transition, the whole mob had broken and was in flight. Men swarmed between the baggage carts, beating and clawing each other in their terror, flinging down swords and daggers as they ran. After them raged the cavalry. The blades swung; victim after victim doubled over, rolled jerking on the grass. The battle became a confused and frantic flight, fanning out across the sloping ground to the rear.

From the tumult rode a man on a grey horse. He moved diagonally towards me, swerving his mount between the clumps and tussocks of grass. After him, at reckless speed, pounded a dozen of the Celts. I saw the gleam of pale hair; then I was riding madly to cut him off. As I rode I yelled, '*Ulfilas* . . . !'

He saw me, but there was no time to check or turn. I struck him at full speed; his horse went down in a flailing of hooves, throwing him headlong. He was up in an instant, coming at me. I swung my shield, heard it connect. He drove for me; I parried and cut, felt the blade bite home. Then it seemed a red mist descended on my sight, so that the men and horses wheeling round were reduced to shadows. His sword flew across the grass, his hand still gripping the hilt. I stood over him and panted, seeing the earth fly, feeling the blade jar through him into the ground. It took a long while for the mist to clear. Then, suddenly, it was gone, and the world was empty.

I stepped back. I said, 'Flog him.'

Riconus stared down at me, his eyes white with killing. He said, 'He's dead.'

'Flog him,' I said. 'He's not dead to me.'

I would, I think, have carried out my scheme; stamped his bones into the earth, and sown the place with salt. But a distant wave of shouting made me turn.

Beyond the battleground the land rose to a wooded ridge. The first of the fugitives, in their stumbling flight, had almost reached it, but none of them gained the cover. As they neared the trees, line after line of men debouched into the field. They carried great targes, starred and strapped with bronze. Their cloaks and leggings were of bright plaid; and each warrior bore a massive sword. The blades glinted briefly; and the Bacaudae crumpled into the grass. Horns bawled from the wood; the ranks paced forward steadily; and behind them, like creatures from another world, I glimpsed the ancient scourge. Black-robed women, their hair wild and long, ran and leaped. Some carried snakes and torches; others, naked, stroked themselves with fire.

I had found King Ossa.

I heard Riconus swear. He galloped from me, beating at his horse. All across the field dazed men turned to stare. The Celts, riding at wild speed, homed on the glittering boar. Behind them, still more or less disciplined, were the lancers from Glevum; the rest boiled into a mob that jostled and was still. The armies faced each other, silently.

I stared. Way off were other riders; strung out across the grass, and coming like the wind. They were moving at right angles, heading for the gap between the forces. I saw the leader, glimpsed her streaming hair.

We met in the dead ground between the ranks. What expression her face wore I won't attempt to say. Finally I turned. A man had ridden forward from the opposing ranks. A barbaric helmet, decorated with boars' tusks and inlaid with gold, covered his face to the chin, but there was no mistaking his voice. He sat a while impassively before he spoke. She turned to me, translating mechanically.

'Tell the Roman this was a mighty killing.'

I waited.

'Ask the Roman, are these the soldiers who will lay waste my land?'

The grass spun suddenly, and flickered.

'Tell him,' I said, 'that they are not. Tell him our business here is done; and that if he wills it, we will leave in peace.'

Another interchange. The slitted eyes watched through the helmet. Finally he nodded.

'Tell the Roman he is free to go; for now he knows I am his master.'

I swallowed. 'Tell the King,' I said, 'the booty of the waggons we leave to him. Also this land; all he can hold, from here to the eastern sea.'

The words were translated. He raised his face to the sky, yelled long and high. The shout was answered from the ranks beyond; a sustained, exultant roar. I turned to Riconus, but I didn't need to speak. Bold fighter though he was, he had already weighed the odds; and slowly, by degrees, the banners dipped, the great boar Standard stooped to touch the earth.

She rode behind me, from the battlefield. The Celts closed round us. Once she paused. She sat her horse silently, looking down. The smoke, dense and evil-smelling, rolled across the grass. Everywhere, round the waggons, between the strakes of their wheels, the bodies lay heaped and still. In the distance a few of the Scoti were already picking over the slain. Nothing else moved across the whole grim field, save one young man, one of the garrison from Corinium. He stood a little way from us, fresh-faced and dark-haired, barely older than Valerius. In his hand he gripped the reddened stub of a sword, and he was crying.

She stared a long time. Then she turned away. She said, 'Were we worth it? Was anything?'

I retired to Corinium.

Riconus came to me in the night. Crearwy lay on my bed, muffled in a cloak, sighing and turning restlessly. I had been keeping vigil by her; I rose at sight of him, and moved away. He spoke so softly his deep voice barely stirred the air. 'This Tribune, from the north,' he said. 'Have you reached a composition with him?'

I said, 'As yet, no.'

He frowned past me at the woman on the bed. He said, 'I don't much like his looks.'

A wait. I said, 'What would you suggest?'

He stared at me. 'The men will ride. We can get new horses.'

I said, 'Tell me when you're ready.'

She rose uncomplaining, drew the cloak round her. We walked through silent streets. The Palatini were drawn up in the shadow of the gates. We burst through in sudden thunder, and there was the straight, long road ahead. Whether we were pursued or not I never knew.

The sky flared behind the towers of Verulam. The gates were closed, men lining the walls. They left us in no doubt as to their intentions. I saw a catapult discharge; the rock hummed over my head, burst on the road behind. I rode forward alone, calling.

'*Pacatianus* . . . *Pacatianus* . . .'

A stirring on the wall; and a question flung itself down to me.

'Who's there?'

'Praefect of Rome.'

'Paullus?'

'Yes . . .'

A wait, and the massive gates squealed back. We swept through, with a clatter and surge.

Wine tinkled into a glass. He handed it to me silently. I drank, feeling the warmth move into my veins. He said, 'The north is in uproar. The Saxon Shore is holding for the Count. His Standards are at Anderita. There's a price on your head, of course. I expect you're aware of that.'

I said, 'You're well informed.'

He rubbed his face. He said, 'We have to be.' His hair, a fringe, was silver, his eyes a bright, direct blue. Now he looked old. He said, 'There was an all-night session. In the Curia. We won our point. The gates stay shut.'

A silence. I swilled the wine round in the glass. He said, 'We could use your men. We shall be fighting the entire Province.'

I shook my head. 'They'll go to Gaul. They won't come back. You'll hold longer than Rome.'

He said, 'It's the end, of course. I find it . . . disappointing.'

There was a statuette on a plinth. A bronze, of Venus. Beyond her the garden was pale with early sunlight. I touched the figure vaguely with one finger. She was coltish and flat-bellied. Britannic.

He said, 'Where will you make for?'

'Rutupiae.'

'Don't. Go to Dubris. Ask for my brother. The quinquennale.'

I put the cup down. I said, 'Will Dubris hold?'

'For a time. Avoid the other garrisons.'

I said, 'I trained them.'

'Yes,' he said. 'I know.'

The port was choked with shipping. Above the town the lighthouse tower thrust gauntly at the sky. A great fire burned on its summit.

There was a merchantman leaving for Gaul in ballast. We embarked the horses. As the last man ran aboard the ropes were cast off, splashed into the harbour. I heard a stir behind us on the quay. Horsemen were coming, moving fast. I saw the swirl of cloaks, the gleam of a sword. Someone hailed us. Nobody bothered to answer.

The night was calm. We edged out through the harbour mouth. As we gathered way a galley, a fighting ship of the British Fleet, came threshing up astern. Riconus said, 'They'll have the place sealed off by dawn.'

I stood beside him, leaning on the rail. A lantern had been hoisted to the masthead. It seemed other vessels followed us, shadows against the pale gleam of the sea. The ship rose and dipped, breasting the first of the swell. I stayed silent, feeling the breeze on my face, hearing the long creak of spars and cordage. Astern the land reduced itself, dropping away like a country seen in a dream. In the west the sky glowed furnace-red, pricked by the orange pinpoint of the tower. As night deepened, massive cloud-streaks grew overhead. Between them, as between the fingers of a hand, the glow still burned. In time the cloud banks thickened, scooping Britannia into the dark.

Chapter twenty

There's not much more to tell.

We didn't stay in Gaul. We rode south, to Hispania. Once over the border we travelled south again. Finally we reached Italica. The place seemed smaller than I remembered it, its streets dusty and narrow. The hooves of the horses echoed from empty house fronts. Signs swung and creaked uneasily, their paint blistered and peeling. There were few people about. Rubbish had collected in the gutters, and there was a smell of drains. I turned into my street, reined outside what had been my home. The shops where Septimius held court, and old Zenobia shouted and pranced, likewise lay deserted. The doors of the house were closed. The man who answered my banging was white-haired and old. It was a moment before I recognised Marcus. The rest of the servants were gone, my father on his death-bed. I was in time to catch his last breath, as he would have wished.

North of the town, the Via Argenta climbs towards the hills. There was a wood, bright with the sounds of birds, a temple crowning a knoll of lush green grass. Within were more birds, many birds, and flowers, painted in gay profusion on the plaster. A spring had been diverted, to feed a fountain. It tinkled and splashed; from it the water flowed in a rivulet down the hill. The tomb I had imagined for Calgaca, my father had built. From here she could see the sky and the hills, the great road lancing to the north. I laid him with her, sat a long time silently after the priest had gone. It seemed for the first time I was beginning to understand him.

Money had been left for me. Not much, but enough for my

needs. We moved south again, to Gades. At the port I commandeered a mailboat. She took us to Tingitana; from there we turned east along the coast. Behind us we left winter.

Such a winter, too, as the West had never known. For weeks on end, for months, icy winds scourged down like the very wrath of God. Belgica, Germania, Britannia, the Gauls, all felt that terrible breath. It froze the sodden earth an arm's-span deep, baking it with ice. It piled snow higher than living memory could record. It choked roads, cut off cities, buried entire towns. It froze birds to the branches on which they huddled, so they dropped off in the spring like feathered stones. It stopped the hearts of old folk and the sick. It burst pipes and wells, aqueducts and sewers, the frames of houses and the sap of forest trees. All this it did, and more.

It froze the Rhine.

The pent-up tide of humanity thus released rolled unchecked, driven by its own colossal momentum, through Belgica and Gaul and Hispania, clear to the sea. Behind it it left ruin, desolation and the dark. While Stilicho lived there was a chance it could be stemmed, even turned back; but he is gone, cut down finally like a dog, as he himself foresaw. No one is left, now, to take his place.

The Court of Mediolanum removed itself to Ravenna; and there I'm told, closed behind unscaleable walls, the Emperor and his minions still play their ghostly games of power. They raise phantom Legions, shadowy Standards, fling back once more the boundaries of Empire, spread, to their hearts' content, the peace and fame of Rome. But there are no Legions, no Standards. The West has fallen.

Roma herself still lives, if a shadow can lay claim to life. It's been many seasons now since Alaric the Goth burst her gates. Twice they bought him off, with pepper and gold, and I hear they raised a petty King or two; but it was food his nation needed more than glory. They rode in, when they came, under the Labarum, Christians and heretics, sworn to respect the property of brothers who disowned them. They didn't understand, they couldn't understand; nobody will ever understand. Neither could they eat stone and bronze. It was a crusade, absurd and holy; and it lasted just four days.

I said Rome lives, but her time can't now be long delayed. The breath comes and goes in her, gusting and harsh, the roar of the amphitheatre; but already, before the body is still, it is attacked by dissolution. In nooks and crannies, sheltered from the streets, the lime-burners are starting their work, stripping the acres and miles of marble. For generations yet, that vast corpse will feed the kilns. Even the Christians wail, but I can feel no tragedy, no grief. For Roma wasn't buildings and statues, temples and baths and libraries. She was people; people who grew tired, as I grew tired, put from them a burden they found crushing. The shell remains, but the spirit has already fled. Had fled, I've come to realise, generations before I was born.

So waste no tears on mortar and brick; pipe your eye though, if you must, for a silly old man with a limp and a stammer and a scruffy ginger wig. Killed not by the fall of Empires, barbarian fury or a thunderbolt of God, but by his own drunken slaves.

My flight, like that of many others, ended in Africa. After months of journeying with Riconus and his men, I once more stood before a Roman governor. He looked me up and down a while before he spoke. 'Well,' he said finally, 'what do you want from me? I suppose with your fine comitatus you'll be looking for a high position in the Army.'

I shook my head. 'No, sir,' I said. 'I want to work on the water supply.'

We have a house here, Crearwy and I; a rambling cavern of a place on the outskirts of what was once Leptis Magna. Houses in Africa tend to be massive and gloomy, their rooms built underground away from the endless heat. In one such room I sit, penning these last few words.

Our life here has been peaceful, and rich. At first I was troubled by recurrent dreams. Sometimes it seemed the pieces of Ulfilas joined themselves, crying some nonsense about a place called Angle-land. At others I would hear what seemed an endless sea of voices, all mingled with the ring and clash of swords; and hands would grip me, hating, drawing me from life and light, down to blood and despair. Sometimes in the nightmare I would cry out; and wake, sweating, to find I lay in Crearwy's arms. In time, the dreams faded; though even now, if I become overtired, it seems I hear a whisper, the last far echo of that din of battle.

From Britannia herself little has been heard. But some information has come through, by this or that devious channel.

Tammonius, that fated man, lived just three months beyond his elevation. Censorinus' second puppet fared no better; for Gratian too, with power once in his hands, refused the dream of Empire. So they elected, finally, a common soldier. His name was Constantine; it was, I suppose, as good a reason as the next to choose an Emperor. Under his leadership the armies of the Province finally crossed to Gaul. There, for a time, they were successful; then the ancient, inevitable pattern swallowed them. Other names, other men, rose for a time, flickered and were extinguished; Gerontius and Constans Caesar, Edobich; and, finally, Maximus. Yet another Maximus; Hispanian, Pretender, last of an unnoble line. Now they are all gone, swept away. In their places, on the thrones and high tribunals, sit Kings with crowns of emerald and iron.

Africa, I think, will hold a few years yet. But for this Province too the end is in sight. Slowly, with the inevitability of Fate itself, the barbarians tighten their grip on the West. They are in Mauretania now. Numidia has fortified her borders; but those walls, like all walls, can be breached. Carthage defies them, but Carthage can be overthrown again. Africa is the lure, and the prize; Africa, once a Granary of Rome. For grain they overthrew the greatest city in the world; and for grain they will come here. Maybe not this year, maybe not the next; but come they must.

Five nights ago, for the first time in many months, I had a dream. It seemed I stood once more within that great house Censorinus owned. The fires still roared, orange-red and dazzling; yet, strangely, the place remained unconsumed. Flowers grew and blossomed despite the heat, birds sang from bright depths of flame. In the dream, figures came walking. I saw Valerius again, with Pelgea at his side. They raised their arms, laughing; and the children too came running, Melinda with her hare-spear, Nessa with her arrows and bow. I would have spoken, certainly I held out my hands; but the vision faded, I woke to a hot, dark room that smelled of spice and sand.

I took it as a portent, and a sign. Perhaps the heart-wounds my pride dealt are healed at last. Later, in my new awareness, I looked at Crearwy. To me she has always remained as I saw her first. Now I saw streaks in her hair that could be silver, lines on her face deeper than those made by Time. This climate will kill her. Last year Marcus died; and we shall stay no longer.

We are going back, to what was once her home. Tomorrow, or the next day, we shall leave. I plan to travel west, retracing my previous route. Beyond Numidia, nobody knows for certain what we shall find. There will be dangers; but I still have the armband given me by Vidimer, I can still clack my few stiff words of German. There are Burgundians both in Hispania and Gaul; we shall have, I think, as good a chance as any. Some trade with Britannia still goes on; we shall follow the coast, try and find a ship bound for the island. If the Gods are generous we shall move north, through and beyond the Wall. There we shall find a tower, reflected in a lake, and the homes of Crearwy's people; and there perhaps, beyond what was the Empire, we shall find rest.

Few of the Celts remain now. Over the years, the blood ties finally dissolved; there was nothing left, anywhere, for which to fight. Some married and raised families; others drifted away, into the great limbo of the West. But Riconus has stayed. He intends to travel with us. The years have streaked his beard with grey; but he remains as jaunty as ever, and is optimistic as to his fate. He has taken a Roman name, a high-flown name, Arturius; he swears he will wed some wild Princess, and breed a race of Kings.

It has come to me of late that I should like a daughter or a son, but Crearwy has remained barren. I think she uses some skill or knowledge from her native land, but I haven't questioned her, and never will. She had a family once; and no man can own the earth. Even the Emperors found that.

This manuscript I shall send, not to Roma, but to Constantinopolis. There at least it may survive for a time; or perhaps, like us, it will vanish. For myself, I have come to a composition with my fate. The childhood dreams of glory are gone, destroyed in that Britannic field. Future ages, certainly, will never hear my name; but it is the here, and the now, that are important. To wake, and sleep, and know that one is loved; to hear, as I hear now, the whisper of a footfall; to feel the touch of hands; these things, above all else, are to be desired. I count myself fortunate in that, for a little while, I have known them.

Time, I think, had bred some tolerance in me; I have become reconciled to many things, even the Christian faith. There is much merit in it; though like Cassianus I misdoubt the future. The Church broke Rome, before Alaric came; and still her power grows. If her dream is realised, she too will found an Empire; and that Empire, over the years, could prove to have in it less mercy, less wisdom and understanding, than the Empire of the Caesars themselves.

I have fallen recently much under the influence of Plotinus and the new followers of Plato. In time, I hope, I may reach understanding; that full understanding without which life is meaningless. I sense, dimly and sporadically, the majesty of the One, Totality, Fulfilment; but my faith and the beliefs I have come to hold would be of no interest here, even were I capable of giving them full expression. It will be enough, perhaps, to speak in the language of metaphor and dreams; to say that, beyond our time and the Lands we know, I have come to feel another may exist. A land of sunshine, and eternal peace. There, perhaps, the children may be said to wait; and there too Crearwy and I may one day travel. For we will build ourselves a home, in that misty country I have never seen, where the red deer shout to the dawn and the heather flows purple to the sea. There we will sit; and there too we will wait. For the Boat of Fate, the White Boat, that Boat that sails for ever.

**SERGUIS PAULLUS BRITANNICUS
WHO WRESTED HAPPINESS FROM TIME
AND FROM THE JEALOUS GODS**

Some place-names of Hispania and Gaul

Barcino	Barcelona
Burdigala	Bordeaux
Carthago Nova	Cartagena
Corduba	Cordoba
Cottian Hills	Alpes Maritimes
Gades	Cadiz
Gesoriacum (Bononia)	Boulogne
Italica	Nr. Santiponce
Massilia	Marseilles
Mediolanum	Milan
Toletum	Toledo
Valentia	Valencia

The place-names of Britannia

Abonae	Sea Mills
Anderita	Pevensey
Aquae Sulis	Bath
Caesaromagus	Chelmsford
Calleva of the Atrebates	Silchester
Camulodunum	Colchester
Corinium	Cirencester
Cunetio	Mildenhall
Deva	Chester
Dubris	Dover
Durobrivae	Rochester
Durocornovium	Wanborough
Durovernum	Canterbury
Eburacum	York
Glevum	Gloucester
Isca of the Dumnonii	Exeter
Isca of the Silures	Caerleon
Lindinis	Ilchester
Lindum	Lincoln
Londinium (Augusta)	London
Luguvalium	Carlisle

Magnis	Carvoran
Manucium	Manchester
Noviomagus	Crayford
Pontes	Staines
Portus Adurni	Portchester
Regulbium	Reculver
Rutupiae	Richborough
Segedenum	Wallsend
Segontium	Caernarvon
Sorviodunum	Old Sarum
Spinis	Speen
Vagniacae	Springhead
Venta	Caerwent
Verlucio	Sandy Lane
Verulamium	St Albans

Her administrative districts

Britannia Prima	Capital at Cirencester
Maxima Caesarensis	Capital at London
Flavia Caesarensis	Capital at Lincoln
Britannia Secunda	Capital at York
Valentia	Capital (probably) at Carlisle

Her rivers

Sabrina Fl. The Severn
Tamesis Fl. The Thames

And her islands

Mona Ins.   Anglesey
Vectis Ins. The Isle of Wight

16